Industrial Clusters in Biotechnology

INDUSTRIAL CLUSTERS IN BIOTECHNOLOGY

Driving Forces, Development Processes and Management Practices

Vittorio Chiesa
Davide Chiaroni
Politecnico di Milano, Italy

ICP
Imperial College Press

Published by

Imperial College Press
57 Shelton Street
Covent Garden
London WC2H 9HE

Distributed by

World Scientific Publishing Co. Pte. Ltd.
5 Toh Tuck Link, Singapore 596224
USA office: 27 Warren Street, Suite 401-402, Hackensack, NJ 07601
UK office: 57 Shelton Street, Covent Garden, London WC2H 9HE

British Library Cataloguing-in-Publication Data
A catalogue record for this book is available from the British Library.

INDUSTRIAL CLUSTERS IN BIOTECHNOLOGY
Driving Forces, Development Processes and Management Practices

ISBN-13 978-1-86094-498-7
ISBN-10 1-86094-498-1

Typeset by Stallion Press
Email: enquiries@stallionpress.com

Printed in Singapore

Preface

This book collects the main results of the *Cleverbio* Project. "*Cluster development and growth in bio-tech: enabling factors and best practices*" (Cleverbio) is a project funded by the *European Commission* within the *Fifth Framework Programme*, within the topic "*Quality of life and management of living resources*". The consortium which carried out the project is composed of six partners: University of Milano-Bicocca (scientific coordinator), Department of Biotechnology and Biosciences, Milano, Italy; Associazione Impresa Politecnico (financial and administrative coordinator), Politecnico di Milano, Milano, Italy; East Region Biotechnology Initiative (ERBI), Cambridge, UK; East Jutland Innovation, Aarhus, Denmark; Heidelberg Technology Park, Heidelberg, Germany; Ecole Superieure d'Ingegnieurs de Marseille, Marseilles, France.

The project studied the cluster development in the biotechnology sector. Clusters can be defined as the geographical concentration of different actors such as interconnected companies, specialised suppliers, service providers, institutions, which compete and cooperate in the same industry. Cluster development is a complex process and usually involves a number of actors such as governmental departments, economic development agencies, public administrations, universities and research centres, companies of different types, financial institutions.

There is a wide body of literature on clusters. However, most works concentrated on the description of the cluster: who takes part to the cluster, his role, how the interactions take place, which are the main advantages of creating and being part of a cluster. Much less attention has been paid to the dynamics of a biotech cluster: how the cluster developed and develops, which are the key factors enabling the cluster to grow, the main problems faced. The project aimed to give an answer to these questions.

The project objective was to define a normative model for cluster development in the biotech sector, which identifies key mechanisms to favour the growth and development of a cluster and the best practices in use to manage a cluster.

To achieve the above objectives the project has carried out:

— an empirical study on biotech clusters, examining how they work and identifying the critical factors enabling the growth and development of a cluster in the biotech sector;
— a detailed analysis of dynamics, triggers, barriers and problems related to a cluster, in order to capture the best practices and provide key recommendations.

The empirical work consisted of the in-depth analysis of the five clusters represented in the consortium, concerning five different European countries: Denmark, Germany, France, Italy, UK. The clusters examined are at different stages of development:

— Cambridge is the most important cluster in Europe and one of the strongest biotech area at worldwide level,
— Heidelberg is a major European cluster and one of the strongest in Germany,
— Aarhus in Denmark as well as Marseille in France are at an early stage of development,
— Milano in Italy is at an embryonic stage of development.

Moreover, other clusters have been analysed, such as Paris-Evry (France), Uppsala (Sweden), Biovalley (Switzerland), Bay Area and San Diego (US) to have a more comprehensive sample.

The ultimate result of the project has been to build a normative model for cluster approach in biotech. The normative model identifies the following aspects:

— pre-requisites to cluster approach, i.e. the conditions which allow a cluster to grow;
— driving forces for cluster growth and development, i.e. key mechanisms enabling the cluster to develop;
— best practices in cluster development and management (in relation to barrier removal, solutions to typical problems to be faced etc.).

The book also shows that biotech clusters born and develop on the basis of different processes: some were born and grew spontaneously thanks to the original co-presence of the key success factors (spontaneous clusters) and some others were born as the result of the actions of public actors. Among the latter, this book shows that different mechanisms and policies were at the origin of the process (industry restructuring policies and industry development policies). Finally in few cases the process of clustering started as a result of a combination of different original conditions (hybrid clusters).

This book therefore intends to be of help for: (i) scholars studying the cluster phenomenon and the process of clustering in the biotech (but also, to a larger extent, in high tech industries); (ii) policy makers, involved in the process of undertaking supporting actions to the development of the biotech sector; (iii) managers of institutions, agencies, initiatives in charge of promoting the development of biotech clusters.

The book is composed of ten chapters. The first chapter provides a brief review of the concept of cluster and gives information about the Cleverbio Project. The second chapter gives an overview on the biotech industry (types of firms, business models, sources of competitive advantage). The chapters from three to seven give an overview on the five cluster examined: Cambridge, Heidelberg, Aarhus, Marseille, Milan, whereas the chapter eight describes the main characteristics of other major biotech clusters in the world. Finally, the chapter nine describes the normative model, showing pre-requisites, driving forces and best practices in biotech clusters, and the chapter ten draws some conclusions, identifying the different development processes and clustering forms in the biotech industry.

On behalf of the consortium, the authors wish to thank the European Commission which made the Cleverbio Project possible and constantly gave its support during the development of the work.

Vittorio Chiesa and Davide Chiaroni
Milano, March 2004

Contributors

Chapter 3

Jeff Solomon
Chief Executive Officer, East Region Biotechnology Initiative (ERBI), Cambridge, UK.

Claire Skentelbery
Business Development Manager, East Region Biotechnology Initiative (ERBI), Cambridge, UK.

Chapter 4

Klaus Plate
Chief Executive Officer, Heidelberg Technology Park, Heidelberg, Germany.

Marion Kronabel
Assistant of the Managing Director, Heidelberg Technology Park, Heidelberg, Germany.

Chapter 5

Gyda Marie Bay
Project Manager, East Jutland Innovation A/S, Aarhus, Denmark.

Jørn Enggaard
Project Manager, East Jutland Innovation A/S, Aarhus, Denmark.

Chapter 6

Jean Laporta
Contracting Professor of Engineering, Ecole Supérieure d'Ingénieurs de Marseille (ESIM), Marseilles, France.

Françoise Perrin
Lecturer in Engineering, Ecole Supérieure d'Ingénieurs de Marseille (ESIM), Marseilles, France.

Authors

Vittorio Chiesa is the ENI Professor of R&D Strategy and Organisation at Politecnico di Milano. He also teaches Business Economics and Organisation at Politecnico di Milano and the University of Milano-Bicocca. He is responsible of the "Technology Strategy" area at the M.B.A. Program of Politecnico di Milano. He obtained his Master Degree in Electronic Engineering *cum laude* at Politecnico di Milano. He was previously with Ciba-Geigy and Pirelli, Senior Researcher at the National Research Council of Italy (Istituto di Tecnologie Industriali e Automazione, Milano), Associate Professor at University of Castellanza, Full Professor at University of Milano-Bicocca. He was Visiting Researcher at London Business School and Guest Professor of Innovation Management at the University of Nijmegen (The Netherlands). He is member of the Board of the Italian Association of Engineering Management. His main research areas concern: R&D management and organisation, technology strategy, international R&D; creation of start-up and spin-off companies; technology transfer. He is author of the books 'R&D Strategy and Organisation', Imperial College Press, 2001; 'La bioindustria', Etas, 2003. He is author of more than 100 papers on international journals, books, conference proceedings.

Davide Chiaroni is a PhD student at Politecnico di Milano, Department of Management Engineering. He achieved his Master Degree in Management Engineering at Politecnico di Milano *cum laude* in 2002. He was previously a Research Assistant at University of Milano-Bicocca, Department of Biotechnology and Biosciences. His research area is strategy and strategic management in high-tech industries.

Contents

Preface v

Contributors ix

Authors xi

1. The Concept of Cluster and the Cleverbio Project 1

    1.1 The Concept of Cluster 1
    1.2 The Advantages from Clustering 3
    1.3 The Cleverbio Project: An Overview 6
    Appendix 9

2. The Biotech Industry: An Overview 17

    2.1 Introduction 17
    2.2 The Bio-Pharmaceutical Value Chain 19
        2.2.1 The Value Chain 19
        2.2.2 Time, Risk and Cost of the Drug Discovery and Development Process 24
    2.3 The Structure of the Bio-Pharmaceutical Industry 26
        2.3.1 Biotech Companies: A Taxonomy 27
        2.3.2 Main Figures in the Bio-Pharmaceutical Industry . 28
    2.4 The Industry Structure: A Geographical Analysis 45

3. The Cluster of Cambridge 49
(*by Jeff Solomon and Claire Skentelbery*)

3.1 History of the Cluster 49
3.2 Major Actors 51
3.2.1 Dedicated Biotech Firms 51
3.2.2 Industrial and Research Environment 57
3.2.3 Financial Environment 61
3.3 Context Factors 63
3.4 Conclusions 66

4. The Cluster of Heidelberg 71
(*by Klaus Plate and Marion Kronabel*)

4.1 History of the Cluster 71
4.2 Major Actors 75
4.2.1 Dedicated Biotech Firms 75
4.2.2 Industrial and Research Environment 82
4.2.3 Financial Environment 84
4.3 Context Factors 84
4.4 Conclusions 87

5. The Cluster of Aarhus 91
(*by Gyda Marie Bay and Jørn Enggaard*)

5.1 History of the Cluster 91
5.2 Major Actors 93
5.2.1 Dedicated Biotech Firms 93
5.2.2 Industrial and Research Environment 99
5.2.3 Financial Environment 101
5.3 Context Factors 102
5.4 Conclusions 103

6. The Cluster of Marseilles 107
(*by Jean Laporta and Françoise Perrin*)

6.1 History of the Cluster 107

6.2 Major Actors 110
6.2.1 Dedicated Biotech Firms 110
6.2.2 Industrial and Research Environment 119
6.2.3 Financial Environment 121
6.3 Context Factors 121
6.4 Conclusions 123

7. The Cluster of Milan 127

7.1 History of the Cluster 127
7.2 Major Actors 129
7.2.1 Dedicated Biotech Firms 129
7.2.2 Industrial and Research Environment 140
7.2.3 Financial Environment 141
7.3 Context Factors 142
7.4 Conclusions 145

8. Other Cases of Biotech Clusters 149

8.1 The Cluster of San Diego 150
8.2 The Bay Area 154
8.3 The Cluster of Evry 157
8.4 The Cluster of Munich 159
8.5 The Cluster of Oxford 160
8.6 The Biovalley 163
8.7 The Cluster of Uppsala 165

9. The Normative Model 167

9.1 Growth Mechanisms of a Cluster 167
9.2 Driving Forces and Practices 169
9.2.1 Financial Driving Forces 170
9.2.2 Scientific Driving Forces 181
9.2.3 Industrial Driving Forces 192
9.2.4 Supporting Driving Forces 201
9.3 The Normative Model 212

10. Conclusions: Forms of Cluster Creation in Biotech 213

10.1 Spontaneous Clusters 214
10.2 Policy-Driven Clusters 215
10.2.1 Industry Restructuring Policies 215
10.2.2 Industry Development Policies 216
10.3 Hybrid Clusters 217

References and Further Readings 219

1 The Concept of Cluster and the Cleverbio Project

1.1 The Concept of Cluster

Marshall (1920) was one of the first economists dealing with the concept of cluster, observing the creation of industrial districts. Marshall noted the apparent importance of industrial localisation while looking at English industrial regions of the 19th century, noticing the intangible dimensions of localisation, as evidenced in his famous comment about the secrets of industry being in the air. Though Marshall made reference to the technological dynamism of English industrial districts, he did not clearly distinguish between localisation as a means of reducing production costs under conditions of market uncertainty and localisation as an underpinning of the technological trajectory of an industry.

In earlier definitions, indeed, geographical concentration was not seen as a major characteristic of a cluster. Czamanski and Ablas (1979) refer to clusters as "a group of industries connected by important flows of goods and services".

Even Porter (1990) in his first contribution to this issue defines an industrial cluster as a set of industries related through buyer-supplier relationships, or by common technologies, common buyers or distribution channels, or common labour pools. Porter provides a simple definition of two types of clusters: vertical clusters and horizontal clusters. Vertical clusters are made up of industries that are linked through buyer-seller relationships, whereas horizontal clusters include industries in which the other kinds of commonalities (market, technology, labour force, ...) prevails. Geographic proximity emphasises advantages of industrial clusters but is not a prerequisite to their identification.

The geographic concentration as key feature in the definition of clusters appears later in the work of Redman (1994): "a cluster is a pronounced geographic concentration of production chains for one product or a range of similar products, as well as linked institutions that influence the competitiveness of these concentrations (e.g. education, infrastructure and research programs)".

Rosenfeld (1995) strengthened in his definition the concept of geographical concentration, identifying a cluster as "a loose, geographically bounded agglomeration of similar, related firms that together are able to achieve synergy. Firms "self-select" into clusters based on their mutual interdependencies in order to increase economic activity and facilitate business transactions".

Jacobs and DeMan (1996) present more in-depth discussions of the different definitions of industry clusters, although these authors also use the original definitions of Porter concerning vertical and horizontal clusters as the basis for their works. Jacobs and DeMan argue that "there is not one correct definition of the cluster concept ... different dimensions are of interest". They expand from the definitions of the vertical and horizontal industry clusters to identify key dimensions that may be used to define clusters. These include: (i) the geographic or spatial clustering of economic activity; (ii) horizontal and vertical relationships between industry sectors; (iii) use of common technology; (iv) the presence of a central actor (i.e., large firm, research centre, etc.); and (v) the quality of the firm network, or firm cooperation. They consider the presence of a central actor as a key feature for a cluster. This represents quite an exception in the literature.

Again Rosenfeld (1997) adds further criteria in defining a cluster including the size of the cluster, the economic or strategic importance of the cluster, the range of products produced or services used, and the use of common inputs. He, however, does not encourage defining clusters exclusively by the size of the constituent industries or the scale of employment, pointing out that many effective clusters are located in small inter-related industries that do not necessarily have pronounced employment concentrations. According to Rosenfeld (1997), an industry cluster is "a geographically bounded concentration of similar, related or complementary businesses, with active channels for business transactions, communications and dialogue, that share specialized infrastructure, labour markets and services, and that are faced with common opportunities and threats". Rosenfeld's definition clearly emphasizes the importance he places on the role of social interaction and firm cooperation in determining the nature of a cluster. Moreover, the latter definition introduces the importance of specialised infrastructures in creating the prerequisite for the establishment of a cluster.

Recent contributions (Porter, 1998; Swann, Prevezer and Stout, 1998; Cooke, 2000; Feser and Bergman, 2000) strengthen the feature of the geographic concentration, assuming a regional perspective to identify clusters.

To summarise, the key features which play a key role in a cluster are: (i) formal input-output relationships; (ii) buyer-seller linkages; (iii) geographic concentration of firms; and (iv) shared specialised infrastructures. Starting from this, in this work, we assumed as definition of cluster the following: "a geographical concentration of actors in vertical and horizontal relationships, showing a clear tendency of co-operating and of sharing their competences, all involved in a localised infrastructure of support".

1.2 The Advantages from Clustering

The definition of cluster itself suggests that clustering may lead to significant advantages for firms. They may take advantage of the strong demand in the location, the large supply of manpower (even high qualified and specialised), and the network of complementary strengths in neighbouring firms. Particularly in high technology industries, geographical proximity plays another pivotal role in the early stages of the life cycle of a product or technology, facilitating the use and transfer of tacit knowledge that is a key to successful development. Two literature contributions related to this issue need to be mentioned.

Porter in his Adam Smith Address (1998) identifies three kinds of advantages in clustering:

(1) *Productivity advantages*: due to the use of better and cheaper specialised inputs (components or services). These come from minimal inventory requirements and lower transaction costs as for the low distance and for the establishment of high trust relations among companies within a cluster. Moreover, joint purchasing services or shared infrastructures (particularly high-tech facilities) may reduce fixed costs for existing companies and initial investments for new ventures;
(2) *Innovation advantages*: proximity between customers and suppliers facilitates the transfer of tacit knowledge. Moreover, the proximity to knowledge centres offers a strong potential for innovation, allowing critical mass to be gained, particularly for pre-competitive activities (for example basic research). Finally, localised benchmarking among actors in the cluster and the great availability of a qualified labour market can strongly improve the capacity to innovate.
(3) *New business advantages*: due to better circulation of information about market opportunities and potential, barriers and risks for new firms can be lower for the clear perception of unfilled needs.

Another analysis on the clustering phenomenon is presented by Swann, Prevezer, and Stout (1998) in their book *The dynamics of Industrial Clustering*. The authors take in account both advantages and disadvantages of clusters, assuming two perspectives: (i) demand side; and (ii) supply side. Table 1.1 shows the results of their analysis.

Concerning the demand side, major advantages are the following:

- *input-output multipliers*: firms located in the same geographic area may take advantage by a strong local demand and/or stimulate induced activities (e.g. dedicated suppliers or services) as well as the demand by other areas, thus creating a virtuous circle that sustains the cluster growth;
- *hotelling*: the term refers to the theory by the economist Harold Hotelling (1929) concerning spatial competition. He provided the empirical evidence that the location of a new firm within a cluster allows to increase its market share thanks to the existence of incumbents;

Table 1.1 Advantages and disadvantages in clustering (source: Swann *et al.*, 1998).

	Demand Side	Supply side
Advantages	• Input-output multipliers • Hotelling • Search costs • Information externalities	• Technology spillovers • Specialised labour • Infrastructures
Disadvantages	• Congestion and competition in output markets	• Congestion and competition in input markets

- *search costs*: the presence of a firm within a cluster may increase its visibility to existent and potential customers allowing them to reducing searching costs;
- *information externalities*: informal relationships favoured by co-location may increase the transfer of tacit knowledge between people working within a cluster;

whereas major disadvantages concern:

- *congestion and competition in output markets*: an increased number of competitors in the same geographic area may reduce, accordingly to microeconomic theories, per-firm sales, prices, profits and growth. These effects, however, actually start to dominate demand side advantages when congestion becomes heavy, suggesting that there may be diminishing (and eventually negative) returns to locating in a cluster as it reaches its maturity.

On the supply side, major advantages are:

- *technology spillovers*: from widespread tacit technology transfer;
- *specialised labour*: the supply of high qualified labour within a cluster is mainly affected by two processes: (i) the ability to generate resources "internally" (favoured by a strong scientific base); and (ii) the ability to attract key people from other geographic areas (related to the visibility of the cluster itself and to the area attractiveness);

- *infrastructures*: the possibility of sharing common facilities, as for Porter, which reduces costs for firms within a cluster.

Disadvantages again refer to congestion and competition in input markets, whether it may be, for example, the cost of real estate or the cost of labour. It is expected that these effects come to dominate for new firms when the cluster reaches its maturity.

It is interesting to notice how both contributions look at clustering as a "spontaneous phenomenon". Possible actions by public actors to increase perceived advantages (e.g. through favourable industrial policies) or to reduce disadvantages (sustaining clusters in their maturity) are not taken into account. Literature contributions concerning the latter aspect, moreover, are rather weak. In most cases, however, particularly in the biotech sector, public interventions actually have been the trigger factor for the birth of clusters.

1.3 The Cleverbio Project: An Overview

This book collects the main results of the Cleverbio Project. "Cluster development and growth in bio-tech: enabling factors and best practices" (Cleverbio) is a project funded by the European Commission within the Fifth Framework Programme, within the topic "Quality of life and management of living resources", Thematic priorities "Research and technological development activities of a generic nature", area "Analysis of social and economic driving forces and of new opportunities in the bioindustries".

The project objective was to define a normative model for cluster approach in the biotech sector, which identifies key mechanisms to favour the growth and development of a cluster and the best practices in use to manage a cluster.

To achieve the above objectives the project has carried out:

- an in-depth study of biotech clusters, examining how they work and identifying the critical factors enabling the growth and development of a cluster in the biotech sector;

- a detailed analysis of dynamics, triggers, barriers and problems related to a cluster, in order to capture the best practices and provide key recommendations.

The concept of cluster is well known and, as we have seen above, there is a wide body of literature. However, most works concentrated on the description of the cluster: who takes part in the cluster, their roles, how the interactions take place, and what are the main advantages of creating and being part of a cluster. Much less attention has been paid to the dynamics of a biotech cluster: how the cluster had developed and continues to develop, which are the key factors enabling at the different stages the cluster to fluorish; and the main problems faced. The project aimed to give an answer to this aspect.

The empirical work consisted of the in-depth analysis of five clusters. They concerned five different countries in Europe: Denmark, Germany, France, Italy, UK. The clusters examined are at different stages of development:

- Cambridge in UK is the most important cluster in Europe and one of the strongest biotech areas at the worldwide level;
- Heidelberg is a major European cluster and one of the strongest in Germany;
- Aarhus in Denmark as well as Marseille in France are at early stages of development;
- Milano in Italy is at an embryonic stage of development but has the potential at both scientific and industrial level to have a strong development in the near future.

Moreover, other clusters have been analysed, such as Paris-Evry (France), Uppsala (Sweden), Biovalley (Switzerland/Germany/France), Bay Area and San Diego (US) to have a more comprehensive sample.

The work allowed to give a description of the cluster looking at the composition of the cluster and the actors taking part to the cluster, the role of each actor, the interactions between the actors, but also how the evolution took place, the main problems to face, and the key decisions taken. The project also examined how biotech clusters have started; what the process of aggregation of the different actors has been; and how the network is working.

Therefore, the project results allow us:

- to compare the development process of the clusters;
- to identify the key stages of development of biotech clusters;
- to make a cross-country comparison at European level of the different working principles of the clusters;
- to make a cross-stage comparison at European level of clusters at different stages of development;
- to find similarities and differences between different cases and finally common practice to cluster approach which could be recommended to other cases.

The ultimate result of the project is therefore a normative model for cluster approach in biotech. The normative model includes the following aspects:

- the pre-requisites to cluster approach, i.e. the conditions which allow the cluster approach to be adopted;
- the driving forces for cluster growth and development, i.e. key mechanisms enabling the cluster to develop (they will be identified appropriate mechanisms in relation to specific phase of development of the cluster and in relation to specific local conditions);
- the best practices in cluster development and management (in relation to barrier removal, solutions to typical problems to be faced, etc.).

The project length has been 30 months. It started in January 2002 and ended in June 2004. The main activities and the milestones of the project are reported below (Table 1.2).

The first activity concerned the development of the framework of analysis and the definition of the methodology by which the clusters on field were examined (see the Appendix). The second activity was to conduct the empirical analysis. The third activity related to the development of a first version of the normative model for biotech cluster development. The fourth phase related to the pilot testing phase, aimed to test the normative model on field in different contexts. This phase led to the revision of the normative model. In this phase, a workshop was held in Heidelberg where

Table 1.2 The Cleverbio Project: activities and milestones.

Main activities	Milestones	Date
Development of the framework and of the methodology of analysis of the clusters	Framework and methodology of analysis	6 months
Field analysis	Report on the individual clusters	12 months
Development of normative model for cluster approach	Normative model	18 months
Pilot testing	Results from pilot applications and revision of the normative model	24 months
Dissemination	Workshop, symposium, book	30 months

representatives of clusters outside the consortium were invited (Uppsala and Paris-Evry). Finally, there was the dissemination phase, including an open symposium and this book.

Appendix

The Framework of Analysis

The framework of analysis used to conduct the empirical survey is summarised in Table A.1.

A detailed description of the framework where each section is further examined is reported in the Table A.2.

Consortium Members

University of Milano-Bicocca, Department of Biotechnology and Biosciences, Milan, Italy

The Department of Biotechnology and Biosciences of the University of Milano-Bicocca is composed of eight full professors, twelve associate

Table A.1 Framework of analysis.

Area	Sub-area
General information on the cluster	
Key organisations	
Major actors	Large companies
	Dedicated Biotech Firms
	Service Companies
	Universities and public research centers
	Financial context
	Incubators and Science parks
Forms of cooperation	Main collaborations
	Forms of industry — industry collaboration
Human resources	
General context factors	Governmental initiatives
	Legal environment
	General acceptance of biotech products
	Economic and financial context
	High-tech industry
	Other organizations and associations
Area attractiveness	
Performance indicators	

professors, ten assistant professors. The Department holds a Master Degree Course in Biotechnology.

It does both research activities and application works related to the application of biotechnology to industry (chemicals) and pharmaceutical (pharmaceuticals and diagnostics). The main areas concern bio-structures, bio-systems, and bio-processes. It also hosts the Centre of Excellence of the Lombardia region aimed to transfer results of biotechnological applications to SMEs in Lombardia. In addition, it is active in areas complementary to the research in hard sciences. Research is done on the management of biotechnological companies, including topics such as: management of research and development projects, financing of start up companies, investment evaluation of R&D projects.

Table A.2 The framework of analysis detailed.

Area	Sub-area	
General information on the cluster		Name or acronym Geographical area Starting year Orientation of the cluster Main fields of application of biotechnology
Key organisation(s)		
Major actors	Large companies	Number
		For each large company: — Sales and revenues — Employees — Core business — Year of establishment in the cluster — R&D expenses — Major biotech products — Supported biotech spin outs — Not supported biotech spin outs
	Dedicated Biotech Firms (DBFs)	Number
		Number and name of public firms
		Number of profitable firms
		For the most important DBFs: — Sales and revenues — R&D expenses — Employees — Year of foundation — Core activity
Forms of cooperations	Main collaborations (intra- and extra-cluster)	University–industry collaborations Industry–industry collaborations
	Forms of industry-industry collaboration (intra- and extra-cluster)	Project funding Alliances Joint ventures Outsourcing

Table A.2 (*Continued*)

Area	Sub-area	
Human resources		Availability of managers Training services and education Intra-cluster mobility Extra-cluster mobility (attractiveness for key people from abroad)
General context factors	Governmental initiatives	Funding to: — Basic research — Applied research — Technology transfer — Cluster development — Entrepreneurship Tax incentives
	Legal environment	Laws and regulations Intellectual property rights policies
	General acceptance of biotech products	
	Economic and financial context	
	High-tech industry	
	Other organizations	
Area attractiveness		Quality of life Access to transport means Traffic jams Availability of general infrastructures Availability of space
Performances indicators		Number of (profitable) companies Number of patents Number of new products Number of potential products in the pipeline Time-to-market Turn over Growth rate of employees

Team members: Vittorio Chiesa (coordinator), Matteo Barberis, Jonathan Brera, Elena Gilardoni, Davide Chiaroni, Simone Prandin, Matteo Conforti, Marco Pasqua.

Associazione Impresa Politecnico, Politecnico di Milano, Milan, Italy

The Associazione Impresa Politecnico (AIP) is a no-profit association founded in 1993 by the Politecnico di Milano aimed at enhancing the relationships among the Politecnico di Milano and companies. The main aims of AIP are seeking for research opportunities, seeking funds for developing research, managing research projects and promoting the related results.

Team member: Alberto Savoldelli (coordinator).

East Region Biotechnology Initiative (ERBI), Cambridge, UK

ERBI is an industry led initiative which was formally started in mid 1997 as an initiative of the local biotech community and local and national government officials. A grant was obtained originally from the DTI (Department of Trade and Industry). Now ERBI raises the vast majority of its financial requirements from private sources.

ERBI's aim is to enhance the growth and development of biotechnology in Cambridge and the East of England, thereby asserting the region as a world-renowned centre of excellence. To this end, ERBI promotes local, national and international networking; supports successful growth of new and emerging ventures, and ensures that the future infrastructure of the region allows seamless growth of the bioscience community.

Team members: Jeff Solomon (coordinator), Claire Skentelbery.

East Jutland Innovation, Aarhus, Denmark

East Jutland Innovation A/S was founded in 1998 by the Ministry of Business and Industry and is located in the Science Park of Aarhus. The shares of the company are owned by the following: the Aarhus University Research Foundation, the Science Park Aarhus A/S, Jyske Bank, INCUBA A/S, the

County of Aarhus, the Municipality of Aarhus, and the Freshwater Centre in Silkeborg.

East Jutland Innovation's mission is to commercialise new ideas and launch new business ventures. Its primary goal is to invest in ideas from researchers, students, and employees working in the research and development departments. Furthermore, East Jutland Innovation is the Technological Transfer Office for the Aarhus University Hospital, Aarhus University, and the Danish Institute of Agricultural Research with a well established network of entrepreneurial businesses.

Team members: Lars Stigel (coordinator), Gyda Bay, Joern Enggaard.

Heidelberg Technology Park, Heidelberg, Germany

The Technology Park of Heidelberg is an International Science Park with focus on Life Sciences. It is located close to the University of Heidelberg and to international research institutes, and acts in strong relations with them in the management of international research projects and the creation of an excellence pole for scientific and technological research and applications.

It works as a centre of a regional network of information and communication. Its main mission is: (i) to co-operate with the government, national and international institutions; and (ii) to co-operate with the ministries and with the major scientific institutes in Heidelberg.

Team members: Klaus Plate (coordinator), Marion Kronabel.

Ecole Superieure d'Ingegnieurs de Marseille (ESIM), Marseilles, France

Groupe ESIM is an establishment of the Chambre de Commerce et d'Industrie de Marseille-Provence, specialising in higher education and research in engineering services for industries. ESIM has 3 schools of engineers, with 750 students, a staff of 120 persons with 65 lecturers and engineers, and 6 departments for research and technology transfer. In addition to scientific core courses, Groupe ESIM gives a special attention to — and develops — management and communication courses. It is also accredited to offer postgraduate courses (such as Technological Resources and Total Quality Management, Large Projects and Programs Management).

As a Chambre de Commerce establishment, Groupe ESIM has permanent relations with industries and specifically with SMEs.

Team members: Jean Laporta (coordinator), Françoise Perrin, Zilè Soilihi.

2 The Biotech Industry: An Overview

2.1 Introduction

Biotechnology, in the 21st Century, is doomed to perform a revolution in many research-based sectors, carrying out a new approach to biological and chemical sciences.

It is currently widely recognised that "modern" biotechnology started in 1953, when Nature published James Watson and Francis Crick's manuscript describing the double helical structure of DNA, the chemical base for genes, proteins, cellular processes, and ultimately life. "Modern" biotechnology can be defined as a collection of technologies, for instance genomics, proteomics, combinatorial biology and chemistry, and high-throughput-screening, which causes a rapid advance in all the traditional life sciences, particularly in the pharmaceutical R&D process.

Since Watson and Crick's work, however, it took more than twenty years (1976) for attending to the birth of the first biotech company (Genentech in San Francisco Bay Area). Another four years were required for the first product on the market: genetically engineered human insulin. In the meantime the US legal context (later followed by other countries) evolved to adapt itself to the biotechnology scientific revolution. In 1980 the US Supreme Court granted the patentability of genetically engineered life forms, and the Congress approved the Bayh–Dole Act allowing scientists from Universities to fully exploit intellectual property rights on their research. In the following years hundreds of science-based companies were created (mainly in US but also in other countries like for instance UK and, more recently, Germany) in order to take advantage of the huge possibilities offered by the new research field.

The possible applications of biotechnology widely vary and concern different sectors:

- knowing the molecular basis of health and disease leads to improved and novel methods for treating and preventing diseases. In the pharmaceutical sector, biotechnology products include quicker and more accurate diagnostic tests, more effective therapies with fewer side-effects because they are based on the body's self-healing capabilities, and new and safer vaccines. A major change will be the advent of the personalised medicine, which offers a one-to-one correspondence between the drug and the individual patient;
- the animal healthcare at the same way is increasingly benefiting from biotechnology. Combining animals and biotechnology results in advances in three major areas: (i) improved animal health products; (ii) advances in human health through studies on animals; and (iii) enhancements to animal products (i.e. modified proteins or enzymes);
- in the agricultural industry, biotechnology can help meet the ever-increasing need by increasing yields, decreasing crop inputs such as water and fertilizer, and providing pest control methods that are more compatible with the environment;
- in the food industry, biotechnology provides new products, lowering costs and improving the microbial processes on which food producers have long relied upon. Many of these impacts will improve the quality, nutritional value and safety of the crop plants and animal products that are the basis of the food industry. In addition, biotechnology offers many ways to improve the processing of those raw materials into final products: natural flavours and colours; new production aids, such as enzymes and emulsifiers; improved starter cultures; more waste treatment options; "greener" manufacturing processes; more options for assessing food safety during the process.

- in the manufacturing industry, biotechnology employs the techniques of modern molecular biology to reduce the environmental impact of industrial processes. Industrial biotechnology also works to make manufacturing processes more efficient for industries such as textiles, paper and pulp, and specialty chemicals. Some observers predict biotechnology will transform the industrial manufacturing sector as much as it has changed the pharmaceutical, agricultural and food sectors. Industrial biotechnology will be a key to achieving industrial and environmental sustainability.

Notwithstanding the variety of possible applications, it is a matter of evidence that currently the pharmaceutical sector is the most affected by biotechnology. On the one hand, biotechnology redrew the R&D and even production processes, allowing scientists to understand and actually exploit the genetic pathways to treat pathologies. On the other hand, biotechnology reshaped the pharmaceutical industry's structure, thanks to the birth of many and different typologies of biotech companies. Therefore, in the rest of the chapter, we provide an overview of the effects of the biotechnology in the pharmaceutical sector.

2.2 The Bio-Pharmaceutical Value Chain

2.2.1 The Value Chain

This section presents a detailed bio-pharmaceutical value chain, highlighting how the technological and scientific improvements allowed by biotechnology changed the traditional primary activities of the sector.

The bio-pharmaceutical value chain (Fig. 2.1) can be divided in two different parts, having as keystone the approval of the new drug by the dedicated public authorities (e.g. Food and Drug Administration (FDA) in US and European Medicine Evaluation Agency (EMEA) in Europe). Pre-approval activities concern the research and development, whereas post-approval activities concern the large-scale production and marketing of a new drug. It clearly appears that the approval of a new drug actually represents the boundary between cash absorption and cash generation. Indeed, only the drugs that respect the standards of efficacy and safety defined by

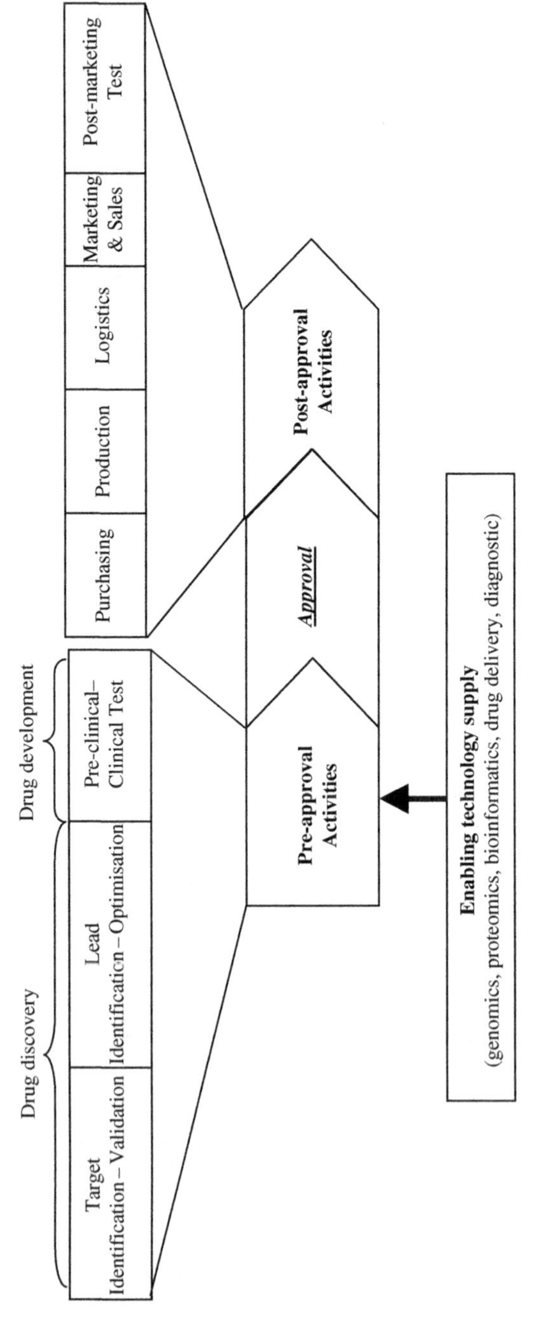

Fig. 2.1: The bio-pharmaceutical value chain.

law can reach the final market, thus generating revenues. It is, however, in the pre-approvals activities that a company creates the conditions necessary to succeed and where, as noted, biotechnology achieves its best results.

Pre-approval activities

Pre-approval activities are the starting point of innovations in the pharmaceutical sector. They include the research of a new chemical or biological compound to be used for therapeutic purposes and the development of the selected ones (leads) into new drugs that can be actually delivered in human beings. The pharmaceutical sector is widely recognised as "research intensive" with the average R&D expenses that were more than 17.5% of the total revenues in the '90s and currently account for nearly 20% in the whole industry. In absolute value, moreover, the R&D expenses had more than doubled since the '90s.

The new research process starts earlier than the traditional one with the target identification. This activity leads to identify a gene or a protein or a sequence of both (target) that is thought to be the pathogenic of a selected disease. The next step is target validation that is concerned with a two-fold study of the identified target. On the one hand, it is necessary to define the interactions between the target and the whole human organism, in order to create a complete disease model that allows scientists to understand the evolution pattern of the disease at the cellular level. On the other hand, the company has to check if there are some intellectual property rights on the selected target (i.e. through the access to public databases as, for example, the NCBI (National Center for Biotechnology Information) in US).

Before biotechnology, these activities did not exist, and the research process was based only on the chemical science and the study of the disease's symptoms.

The same pattern showed for the target has to be followed by the chemical or biological compound. After the genetic base of the disease evolution is known, indeed, scientists need to identify a set of compounds (lead identification) thought to have the desired effects in treating the selected disease. Through several different tests, both *in vitro* and *in vivo* (and recently also *in silico*, using computer aided drug design programs), scientists select the compound that shows the best results in the defined set. This compound actually represents the active principle of the future drug. The lead optimisation, finally, adds to the lead the necessary excipients (i.e. substances included in the drug formulation in order to protect, support or enhance the stability of the active principle and to increase patient compliance).

The final output of the research activities, as defined here, is therefore a drug "prototype" or candidate that can actually enter the development phase, that will study (test) the efficacy and the safety of the new drugs. The first development activity is the carrying out of pre-clinical tests. The scope of these tests is to evaluate, particularly through *in vivo* testing, the effects of the drug on animals. In particular, the mechanisms of absorption, distribution, metabolism, excretion and toxicology (wholly referred to with the ADMET acronym) are studied, in order to effectively trace the pattern of the new drug's compound inside the organism and eventually make some corrections before entering human tests. If only the complete results of pre-clinical tests are positive, then the development phase of the new drug continues with the clinical test activity. Before entering clinical trials, however, a first approval by the public authorities is required. In the US, for example, an Investigational New Drug Application (IND) is needed for the authorisation from the FDA before a new drug can be administered to humans. Such authorisation must be secured prior to interstate shipment and administration of any new drug. Similar authorisations are required in Europe and in the other nations. Clinical tests directly involve human patients and are usually divided into three steps: Phase I, Phase II and Phase III. Some facts about clinical trials can be summarised in the table below (Table 2.1).

In Phase I, researchers test the new drug in a small group of people (20–80) to evaluate its safety, determine a safe dosage range, and

Table 2.1 Clinical tests (*source*: FDA, 2002).

Phase	Number of patients	Length	Purpose
I	20–80	Several months	Mainly safety
II	100–300	Several months to two years	Some short-term safety but mainly effectiveness
III	1,000–3,000	One to four years	Safety, dosage and effectiveness

identify side effects. It may include healthy participants and/or patients. In Phase II, the new drug is given to a larger group of people (100–300) to evaluate the effectiveness of the drug for the particular indication in patients with the disease and to determine the common short-term side effects and risks. Finally, the Phase III involves an even larger group of people (1,000–3,000) to confirm the effectiveness of the new drug, to evaluate the drug's overall benefit-risk relationship, and to provide and adequate basis for physician labelling. If all three phases are successful, the specific public authority approves the new drug that can reach the market.

Post-approval activities

The post-approval activities are quite common to almost all industrial companies and, as previously noted, are concerned with the large-scale production and marketing of the new drug. Hence, they can be identified as follows: (i) purchasing; (ii) production; (iii) logistics; (iv) marketing and sales; and (v) post-marketing test. Some of these clearly do not require a full description. A few remarks are needed, however, particularly about production and post-marketing test.

The production activity, indeed, in the bio-pharmaceutical sector almost always requires complex process engineering, particularly because of the relative novelty of biotech technologies and their differences with the traditional chemical-based ones. Moreover, biological compounds which are on the average greater in molecular size than the chemical ones, are more difficult to manage in the industrial processes (e.g. the compounds' stability is a critical variable).

Finally, post-marketing tests (that some authors call alternatively Phase IV of the clinical tests) delineate additional information including the drug's risks, benefits, and optimal use in the middle-term. This monitoring, which is actually stronger in the period close to the market launch of a new drug, continues for the whole life cycle of the product, thus ensuring its safety also in the long-term horizon.

Enabling technology supply

Support activities, particularly the ones concerned with the technological and scientific base, are essential to reach the final goal of discovering and developing a new drug. Enabling technology supply refers to the activities aimed at generating scientific and technological improvements through both basic research and technological devices' engineering.

Main examples of the scientific and technological applications are the followings: (i) genomics and proteomics, which are the scientific study of genes (or proteins) and of their roles in the organism's structure; (ii) pharmacogenomics, which deals with the specific differences in response of living structures due to the different genome of individuals; and (iii) bioinformatics, which is the creation (i.e. through physical devices), collection, storage, and efficient utilisation (i.e. through software programmes) of biological data.

2.2.2 Time, Risk and Cost of the Drug Discovery and Development Process

The drug discovery and development process, analysed in the previous sections, requires a long time horizon (from 10 to 15 years). During the '90s, moreover, also as a consequence of the use of biotechnology, the time to gain the approval for a new drug significantly increased. Figure 2.2 shows, particularly, the contribution of each phase to the overall completion time.

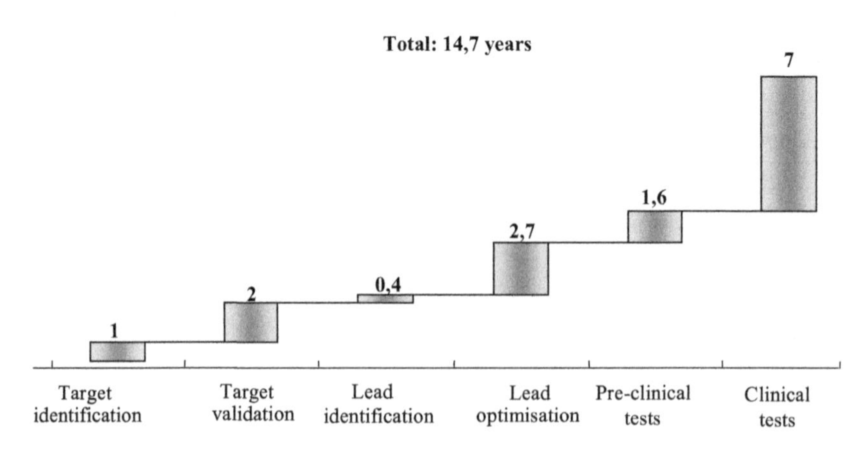

Fig. 2.2: Years to complete each phase in the drug discovery and development process (*source*: The Boston Consulting Group, 2001; IBM, 2003).

The biotech revolution is the major cause of the increase of time in R&D process for at least two reasons: (i) biotechnology allows the treatment of more complex pathologies (as, for instance, chronic and degenerative diseases) that require much more time in clinical tests to verify the effectiveness of therapies; and (ii) biotechnology is still quite far from maturity and technological improvements require a constant revision of the development process.

Figure 2.3 shows that even the approval time (i.e. the time the approval authorities need to examine and validate the results of clinical tests) increased in the last years, mostly because of the weaknesses of rules concerning new biotechnological therapies.

Risk is another key feature of the drug discovery and development process. Only 250 out of 5,000 compounds identified in research actually enter the pre-clinical tests, and only 1 out of 5 drugs passes the clinical tests (Table 2.2). Advanced screening technologies help companies in increasing the examination rate of new compounds, but currently have little effects on the failure rate. Obviously, this is a consequence of the above mentioned distance from technological maturity of such devices: in the future, indeed, they are doomed to strongly reduce failures in the sector.

As an effect both of time and risk, also costs of the R&D process for a new drug are significantly high. Recent studies reported an increase in

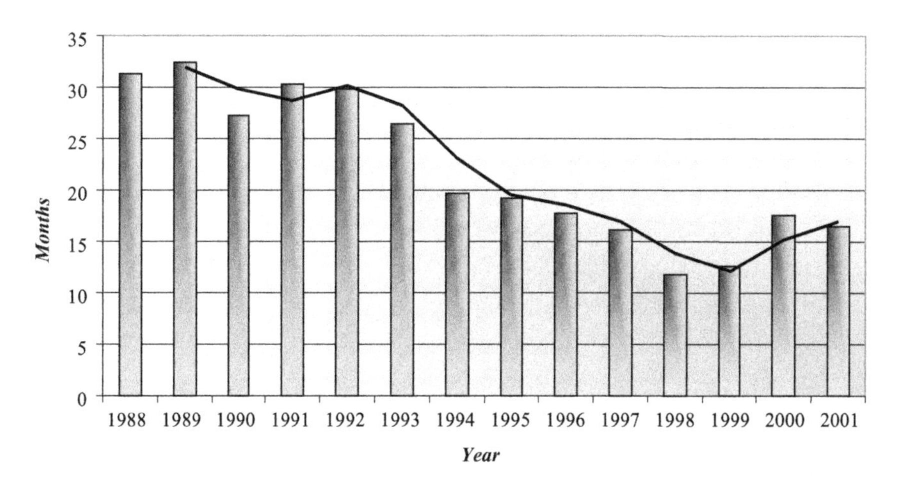

Fig. 2.3: Mean approval time at FDA (*source*: PhRMA, 2002).

Table 2.2 Success rate for each phase in the drug discovery and development process (*source*: Ernst and Young, 2000).

Identified leads	5,000–10,000
Candidates entering pre-clinical tests	250
Drugs entering clinical tests	5
Phase I	80% pass
Phase II	30% pass
Phase III	80% pass
Drugs marketed	1

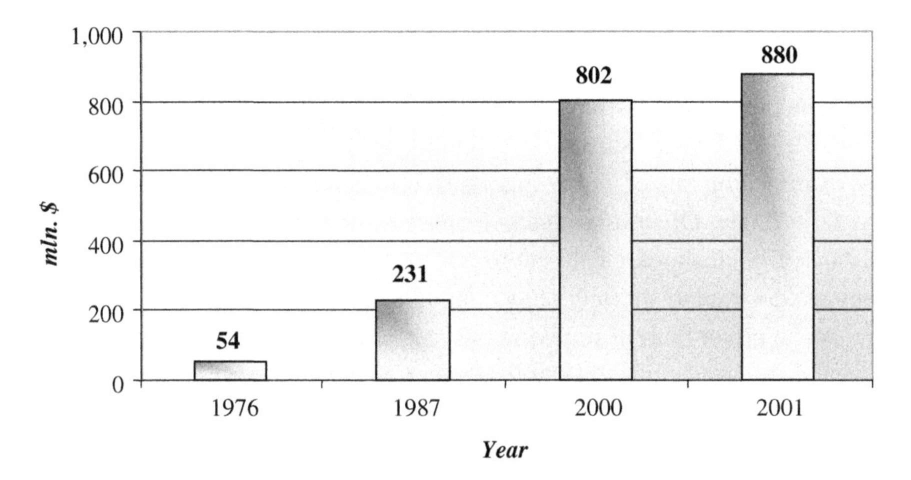

Fig. 2.4: Average development cost for a new drug (*source*: PhRMA, 2002; DiMasi, 2001).

the average cost of development of nearly 280% in the last decade: from US$231 million in 1987 to US$880 million in 2001 (Fig. 2.4).

It is to notice that in Fig. 2.4 the risk-adjusted cost (i.e. the cost of a successful drug and of the associated failures in previous phases) is considered.

2.3 The Structure of the Bio-Pharmaceutical Industry

The industrial structure of the pharmaceutical sector presents a first major distinction between "traditional" large pharmaceutical companies and biotech companies:

- large traditional pharmaceutical companies (the so-called Big Pharma like Pfizer, Johnson & Johnson, Merck, GlaxoSmithKline, . . .);
- biotech companies (the first one was Genentech in 1976), basing their business on the exploitation of the results of application of the new scientific disciplines (genomics, proteomics, ...) based on biotechnology.

2.3.1 Biotech Companies: a Taxonomy

The biotechnology companies (the so-called Dedicated Biotech Firms — DBF) in the pharmaceutical sector may be divided in two major typologies (Fig. 2.5):

- *core biotech companies*, i.e. companies developing (and in some cases commercialising) new products and/or technologies directly related to the drug discovery and development process;
- *complementary product/service suppliers*, i.e. companies offering complementary products (e.g. basic chemicals and reagents) or support services.

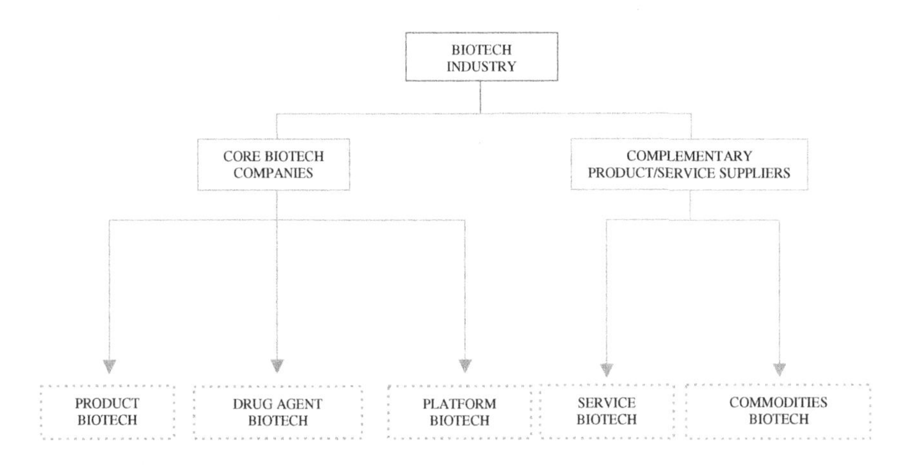

Fig. 2.5: The taxonomy of biotech companies.

Core biotech companies can be further divided into:

- *Product Biotechs*: these companies have as objectives the discovery, development, production and commercialisation of therapeutics products, covering all the activities of the value chain;
- *Drug Agent Biotechs*: these companies develop active principles for new drugs and usually out licence the exploitation rights of their "intermediate products" to larger companies, both Big Pharmas and Product Biotechs;
- *Platform Biotechs*: these companies focus on the development and commercialisation of new technologies and devices, both software and hardware, used in the drug discovery and development process.

Complementary product/service suppliers, moreover, can be further divided into:

- *Service Biotechs*: these companies offers consultancy as well as support services to companies conducting the drug development process;
- *Commodities Biotechs*: these companies act as suppliers of specific chemical provisions, often developed accordingly with the "instructions" of pharmaceutical companies.

2.3.2 Main Figures in the Bio-Pharmaceutical Industry

This section deals with the analysis of each typology of companies, both pharmas and biotechs. Particularly, the following aspects have been taken into account: (i) the business model; (ii) the positioning on the value chain; and (iii) the list of top ten companies by market capitalisation at worldwide level.

Big pharmas

Large pharmaceutical companies still rely upon an absolute market leadership position in the sector. The biotechnology revolution, however, actually weakened their competitive position as it strongly affected the competences required in the drug discovery process. New technologies, like the high throughput screening technologies (HTS) and research approaches based on

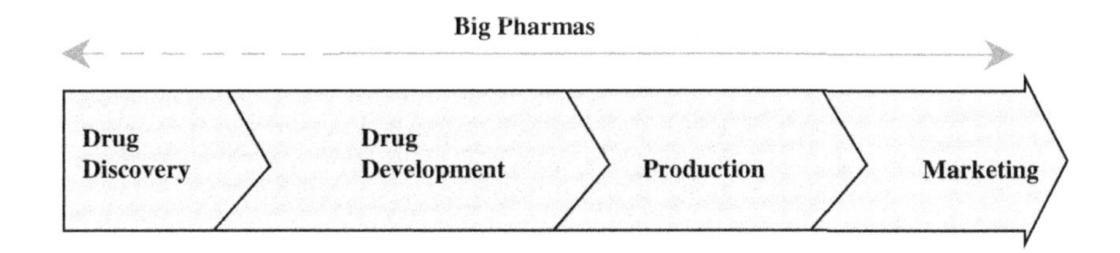

Fig. 2.6: The positioning of Big Pharmas on the value chain.

genomics and proteomics, allow the development of more effective drugs and increased efficiency in R&D. Big Pharmas had to change their business model.

They have progressively specialised downstream in the value chain (drug development, production, marketing), whereas upstream activities (drug discovery) have been the field of specialisation of biotechs (Fig. 2.6). Early phases of the R&D process are covered through: (i) research agreements with innovative biotech companies; (ii) licensing agreements; and (iii) direct acquisition of small biotech companies.

The strength in the post-approval activities actually represents the source of the competitive advantages of Big Pharmas. In particular, the sales force and the garrison on distribution channels allow the pharmaceutical companies to control the market.

Top ten pharmaceutical companies in the world are presented in Table 2.3. Economic fundamentals, and particularly net earnings, are largely positive. The ratio between revenues and R&D expenses (0.15) is reducing (0.18 in 2000 and 1999, and 0.20 in 1998).

This provides the evidence of the above mentioned trend to use "external sources" to replace internal basic research activities. If the pipeline (i.e. the products in development) is considered, only few products are currently in pre-clinical phases: the large majority of the products are in development and this is the result of licensing agreements or acquisitions.

Product Biotechs

Product Biotechs actually play a major role within biotech companies: through their direct action the scientific innovation is "transferred" into

Table 2.3 Top ten pharmaceutical companies by market capitalisation.

Company	Market	Market cap 28/03/2003 (*mln. $*)	Revenues 2002 (*mln. $*)	R&D expenses (*mln. $*)	Net income 2002 (*mln. $*)	Pipeline		Therapeutic areas
Pfizer*	NYSE	195,948.0	32,373.0	4,800.0	9,126.0	Pre-clinical Clinical Filing	24 46 10	Mental health, brain and nerve disorders
Johnson & Johnson	NYSE	170,417.0	36,298.0	1,100.0	6,597.0	Pre-clinical Clinical Filing	0 8 4	Infectious diseases, brain and nerve disorders
Merck	NYSE	124,291.5	51,709.3	2,400.0	7,149.5	Pre-clinical Clinical Filing	14 18 3	Immune disorders, infectious disease
GlaxoSmithKline	NYSE	107,092.3	33,258.3	3,800.0	6,492.6	Pre-clinical Clinical Filling	0 53 14	Infectious diseases, immune disorders
Novartis	NYSE	106,135.9	23,606.5	2,200.0	5,631.0	Pre-clinical Clinical Filing	2 36 8	Cancer, immune disorders

Table 2.3 *(Continued)*

Company	Market	Market cap 28/03/2003 *(mln.$)*	Revenues 2002 *(mln. $)*	R&D expenses *(mln. $)*	Net income 2002 *(mln. $)*	Pipeline		Therapeutic areas
Lilly	NYSE	65,519.4	11,077.5	2,200.0	2,707.9	Pre-clinical	0	Cancer,
						Clinical	28	hormonal
						Filing	5	disorders
AstraZeneca	NYSE	59,117.2	17,841.0	2,700.0	2,836.0	Pre-clinical	2	Brain and
						Clinical	31	nerve
						Filing	6	disorders, cancer
Roche	SWX	58,392.8	21,649.4	1,300.0	−3,100.2	Pre-clinical	10	Cancer,
						Clinical	26	immune
						Filing	2	disorders
Wyeth	NYSE	54,106.7	14,584.0	1,800.0	4,447.2	Pre-clinical	5	Cancer,
						Clinical	13	infectious
						Filling	7	diseases
Bristol Myers Squibb	NYSE	41,781.2	18,119.0	1,900.0	2,066.0	Pre-clinical	16	Cancer, heart
						Clinical	35	and blood
						Filing	9	vessel disorders

*Before the acquisition of Pharmacia the 16th April 2003.

new therapeutics. The end market for these products comprises two major segments:

- *primary care segment*, concerning the products actually and directly available for the end user (the patient);
- *secondary care segment*, concerning the products that require the direct interaction between patients and specialised physicians for their delivery. It is the case, for example, of the new gene or stem cell therapy.

Particularly in the secondary care segment, biotechnology had a major impact because it allows entirely new therapies based upon genetic engineering products to be defined. This means that not only did biotechnology increase the efficacy of traditional drugs but it also paved the way for a new concept of drug product, thus differentiating significantly Product Biotechs from traditional pharmas.

Currently, Product Biotechs, most of which founded in the '80s, show a relevant size. Starting from a strong focus on pre-approval activities and also facilitated by the novelty of biotech applications in the pharmaceutical sector, they rapidly developed reaching the final market through direct commercialisation at global level.

The positioning on the value chain (Fig. 2.7) shows how these companies carry out internally all the activities, from basic research to marketing. Despite the high risks and investments in R&D (US$212 million per company on the average, accounting for nearly 30% of revenues), reaching the market with a new drug is equal to cross the border between cash absorption and cash generation. The average positive profitability is a key characteristic of Product Biotechs, in comparison to the other kinds of biotech companies analysed. Such business model, indeed, can be referred

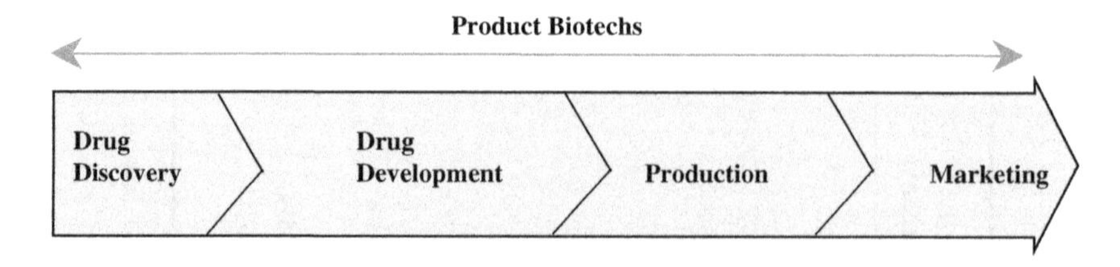

Fig. 2.7: The positioning of Product Biotechs on the value chain.

to as a "virtuous" circle in which the products already on the market fund the research projects, thus contributing to generate a "thick" stream of potential new drugs through the pipeline and to enhance the long-term profitability.

In order to enhance this process, Product Biotechs:

- differentiate their product portfolio, particularly in personalised medicines, thus maintaining a competitive advantage on traditional large pharmas. Biotech drugs, indeed, generally win a premium price in comparison with the traditional ones;
- cooperate with Platform Biotechs: on the one hand, to develop more efficient and effective technologies that reduces risk in the drug discovery process; on the other hand, to implement diagnostic devices for personalised medicines (thus strengthening differentiation effects);
- sustain their growth through acquisitions of smaller companies.

The top ten Product Biotechs are listed in the Table 2.4. The following aspects can be highlighted:

- R&D expenses account for nearly 30% of revenues, showing that the research and development process represents the key activity for these companies;
- the number of biotech drugs (160) directly marketed by the top ten Product Biotechs represents a significant part of the biotech products (nearly 300) on the market.

Drug Agent Biotechs

Differently from the Product Biotechs, Drug Agent Biotechs carry out only in part the drug discovery and development process, usually up to the lead identification and validation (i.e. the identification of a chemical or biological compound with potential therapeutic effects). Therefore, their output represents an "intermediate product", that requires a further development phase (even longer and more risky than the previous ones). As a consequence, Drug Agent Biotechs operate in the industrial market and usually have as customers larger companies (either traditional pharmas or Product

Table 2.4 Top ten Product Biotechs by market capitalisation.

Company	Market	Market cap *(mln. $)*	Revenues *(mln. $)*	R&D expenses *(mln. $)*	Net income *(mln. $)*	Employees	Pipeline	
Amgen Inc.	Nasdaq	62,216.80	5,523.00	1,116.00	(1,391.00)	7,700	Pre-clinical	9
							Clinical	35
							Marketed	34
Genentech Inc.	Nasdaq	17,067.30	2,252.30	623.50	63.80	5,200	Pre-clinical	16
							Clinical	54
							Marketed	32
Serono SA ADS	Nasdaq	12,295.40	1,546.50	342.70	320.80	4,501	Pre-clinical	13
							Clinical	21
							Marketed	10
CHIRON Corp.	Nasdaq	7,073.40	972.90	325.80	180.80	3,736	Pre-clinical	4
							Clinical	29
							Marketed	15
Medimmune Inc.	Nasdaq	6,820.00	847.70	144.20	(1,098.00)	1,600	Pre-clinical	11
							Clinical	16
							Marketed	6
Gilead Sciences Inc.	Nasdaq	6,687.20	446.80	144.30	72.10	1,100	Pre-clinical	0
							Clinical	8
							Marketed	10

Table 2.4 *(Continued)*

Company	Market	Market cap *(mln. $)*	Revenues *(mln. $)*	R&D expenses *(mln. $)*	Net income *(mln. $)*	Employees	Pipeline	
Genzyme Corporation	Nasdaq	6,344.70	1,329.50	308.10	85.00	5,500	Pre-clinical	4
							Clinical	14
							Marketed	16
Biogen Inc.	Nasdaq	5,971.50	1,148.40	360.10	238.70	1,992	Pre-clinical	5
							Clinical	24
							Marketed	10
Idec Pharmaceuticals Corporation	Nasdaq	5,076.10	404.20	90.00	148.10	692	Pre-clinical	0
							Clinical	12
							Marketed	14
Cephalon Inc.	Nasdaq	2,685.00	506.90	116.40	175.10	300	Pre-clinical	2
							Clinical	10
							Marketed	13

Biotechs), actually completing the development process and, if successful, marketing the drug.

The current business model of Drug Agent Biotechs is based on generating returns from the licensing agreements (royalties and similar payments) that typically occur after the lead optimisation. In some cases, the development process of the new drugs is carried out up to clinical tests, where the products reach the higher "transactional value" before commercialisation.

Therefore, Drug Agent Biotechs cover only part of the value chain (Fig. 2.8). Returns are much lower than Product Biotechs comparatively, and the average profitability is yet usually negative. In order to avoid, especially in the current negative economic context, the settlement of a "vicious" circle in which few financial resources result in few new research projects, other than the mentioned licensing strategies, Drug Agent Biotechs:

- exploit externalities in their products portfolio, searching for possible side applications in fields other than the pharmaceutical one (e.g. the use of enzymes for the industrial bio catalysis or the use of active principles in animal healthcare). This allows to increase the amount of royalties related to a single product;
- diversify their business in platform technologies development, thus coupling increased efficiency and efficacy in internal research with increased revenues from supply to other companies.

The top ten Drug Agent Biotechs are listed in the Table 2.5. The greatest ratio among biotech companies between R&D expenses and revenues (more than 100%) is, rather than the outcome of an over-investing trend, the negative effect of the poor stream of returns.

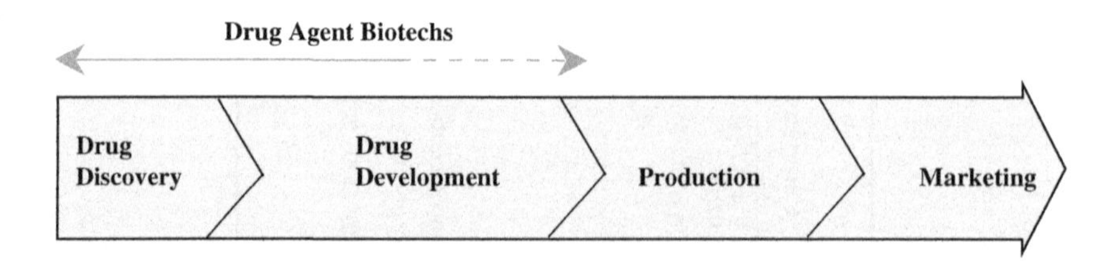

Fig. 2.8: The positioning of Drug Agent Biotechs on the value chain.

Table 2.5 Top ten Drug Agent Biotechs by market capitalisation.

Company	Market	Market cap (mln. $)	Revenues (mln. $)	R&D expenses (mln. $)	Net income (mln. $)	Pipeline	
Trimeris Inc.	Nasdaq	921.30	1.10	51.20	(75.70)	Pre-clinical	0
						Clinical	5
						Marketed	0
NPS Pharmaceuticals Inc.	Nasdaq	766.50	2.20	80.10	(86.80)	Pre-clinical	1
						Clinical	7
						Marketed	0
Abgenix Inc.	Nasdaq	645.40	19.30	124.40	(208.90)	Pre-clinical	4
						Clinical	7
						Marketed	0
Genta Inc.	Nasdaq	562.50	3.60	48.10	(74.50)	Pre-clinical	4
						Clinical	18
						Marketed	1
CV Therapeutics	Nasdaq	493.80	5.30	98.70	(107.80)	Pre-clinical	2
						Clinical	5
						Marketed	0
Telik Inc.	Nasdaq	413.50	1.30	26.60	(34.80)	Pre-clinical	3
						Clinical	8
						Marketed	0

Table 2.5 (*Continued*)

Company	Market	Market cap (*mln. $*)	Revenues (*mln. $*)	R&D expenses (*mln. $*)	Net income (*mln. $*)	Pipeline	
Tularik Inc.	Nasdaq	409.50	25.30	107.20	(93.80)	Pre-clinical	1
						Clinical	19
						Marketed	0
Tanox Inc.	Nasdaq	400.90	0.50	22.70	(26.00)	Pre-clinical	2
						Clinical	4
						Marketed	0
Medarex Inc.	Nasdaq	301.70	39.50	82.60	(157.50)	Pre-clinical	3
						Clinical	19
						Marketed	1
Inspire Pharmaceuticals Inc.	Nasdaq	241.40	4.90	24.80	(24.70)	Pre-clinical	0
						Clinical	6
						Marketed	1

Platform Biotechs

Platform Biotechs represent a complex set of companies based on different technologies that operate as enabling suppliers for core biotech companies. In order to better analyse their role in the sector, a further distinction has to be made concerning the major focus of their "products" between:

- *process technologies*, which improve the efficacy and effectiveness of the drug discovery and development process (e.g. reducing the failure rate or increasing the speed of the analysis); and
- *product technologies*, which are either embedded in the product as, for example, new delivery systems or associated to the product as the diagnostic tools.

This leads to define four typologies of Platform Biotechs (Fig. 2.9):

- x-Omics Platforms;
- Bioinformatics

developing process technologies and

- Drug Delivery Platforms;
- Diagnostics Platforms

developing product technologies.

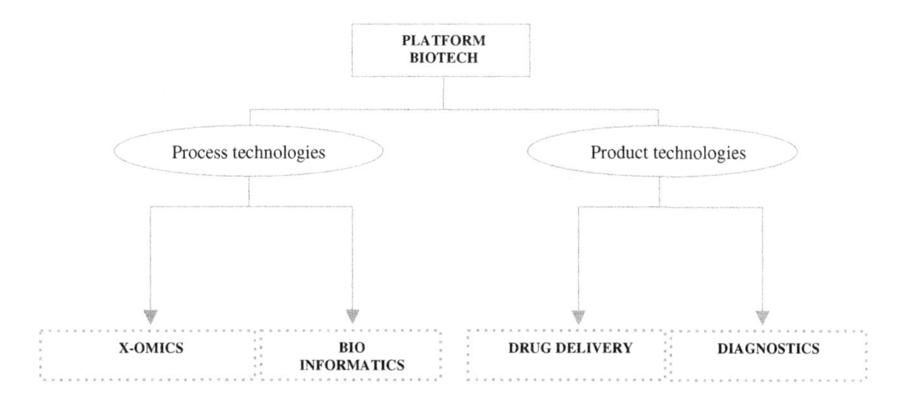

Fig. 2.9: Platform Biotechs.

(i) *x-Omics Platforms*

x-Omics Platforms intervene in the basic research activities, developing and commercialising technological tools and devices to support the drug discovery process. Major examples are Genomics and Proteomics Platforms that develop dedicated systems for the functional and structural analysis of genes and proteins as triggers of the cellular mechanisms. Besides these, other relevant platform technologies are under development: (i) phenomics, which analyses the structure and the visible characteristics of micro-organisms; and (ii) metabolomics, which analyses the effects on the cellular mechanisms of external stimuli or of a genetic mutation. Therefore, it is possible to say, accordingly to the large majority of scholars and practitioners, that biotechnology entered the omic-era.

(ii) *Bioinformatics Platforms*

Bioinformatics Platforms develop and implement tools, both software and hardware, to generate, collect and manage data in output by the drug discovery process. Bioinformatics Platforms act at different stages of the bio-pharmaceutical value chain (form basic research to clinical tests), their major contribution being in the creation of a knowledge base for the whole drug discovery and development process.

Here we consider as bioinformatics technologies both the ones based upon scientific disciplines like combinatorial chemistry and biology (e.g. high throughput screening) as examples of process "automation", and the data storage and data mining technologies as examples of data management tools. The widespread use of biotechnology in the pharmaceutical sector increases the need for informatics technologies able to manage a large amount of data.

(iii) *Drug Delivery Platforms*

Drug Delivery Platforms support specifically the candidates' development, developing and commercialising systems and devices to effectively delivery active principles in human bodies. The delivery of a drug represents a critical phase as it is responsible for a large part of the effectiveness of the therapy and it is often a cause of adverse reactions. Biological molecules (that are the base of new biotechnological drugs) are more complex to manage in industrial processes than the traditional chemical ones, and also their interaction with human bodies are rather difficult to control. Moreover, improvements in genetics lead to identify the specific cellular target for each drug, thus claiming for systems able to deliver "the right drug at the right place". Drug Delivery Platforms play their role in the sector actually on two sides: (i) improving the characteristics of traditional delivery systems (oral and nasal); and (ii) creating completely new systems. Among the latter, it is possible to highlight the lyposomial technologies. Liposomes are microscopic lipid spheres used to encapsulate and deliver the active principles to areas of disease within the body.

(iv) *Diagnostics Platforms*

Finally, Diagnostics Platforms focus on the design and development of devices able to identify, at cellular level, the stages of development of a defined pathology and eventually the genetic susceptibility of each individual to come down with it. Diagnostics Platforms intervene primarily during clinical tests and, if the related drug is successful, their products are further commercialised in the final market as a complement to drugs.

These companies, started recently as a major consequence of the improvements in pharmacogenomics, currently still play a marginal role in the sector. Given the strong linkages between the development of diagnostic tools and of personalised medicines however, most practioners look at Diagnostics Platforms as the companies with the high growth potential in the near future.

Figure 2.10 shows the positioning on the value chain of each typology of Platforms Biotechs.

Platform Biotechs have a time-to-market and a failure rate consistently lower than the drug-oriented companies (Product Biotechs and Drug Agent Biotechs). Platform Biotechs are, therefore, much more similar to ICT companies rather than to "drug-oriented" biotechs, even if they share almost all the needed competences with the latter.

In order to sustain their business model, Platform Biotechs have:

- to develop customised products, able to fit the needs of drug oriented companies, maximising efficiency and effectiveness in different phases of the value chain and in different therapeutic areas;

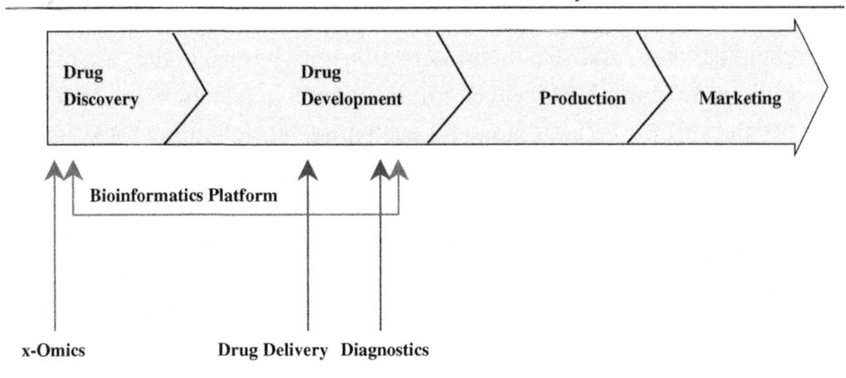

Fig. 2.10: The positioning of Platform Biotechs on the value chain.

- to force the creation of a technological standard (particularly for x-Omics and Bioinformatics companies), thus allowing companies to compete on the whole market and avoiding in house solutions from core biotech companies.

Platform Biotechs, on the average, are not yet profitable but in comparison with Drug Agent Biotechs (with whom they share the same average net losses and the years of birth), they present three times higher revenues and market capitalisations. However, Platform Biotechs are expected to be unable to generate and sustain long-term profits, and in particular, almost all the practitioners agree that they will not reach the results of top Product companies with their "pure", technologically-focused business model.

The top ten Platform Biotechs (Table 2.6) include examples of each typology.

Service and Commodities Biotechs

Service and Commodities Biotechs play a marginal role in the bio-pharmaceutical sector. They present a less interesting case in the analysis of business models, being characterised by a very low risk/reward profile and by investments not comparable with the ones of the other biotech companies.

Service Biotechs actually act as external "source" of information and competences particularly for drug oriented biotech companies (Product and

Table 2.6 Top ten Platform Biotechs by market capitalisation.

Company	Market	Market cap (*mln. $*)	Revenues (*mln. $*)	R&D expenses (*mln. $*)	Net income (*mln. $*)	Typology
Quest Diagnostic Inc.	Nasdaq	5,564.60	4,108.10	n.d	322.20	Diagnostics
Biovail Corp.	Nasdaq	4,129.80	788.00	53.70	87.80	Drug Delivery
Applied Biosystem Group	Nasdaq	3,669.60	1,666.90	234.90	134.30	Bioinformatics
Affymetrix Inc.	Nasdaq	1,336.30	289.90	69.50	7.90	Bioinformatics
Human Genome Sciences Inc.	Nasdaq	1,134.30	3.60	182.60	(219.70)	x-Omics Oriented
Diagnostic Products Corp.	Nasdaq	1,102.40	324.10	34.70	47.30	Diagnostics
Andrx Corp.	Nasdaq	1,041.00	771.00	46.20	(86.40)	Drug Delivery
IGEN International Inc.	Nasdaq	1,016.80	45.30	24.50	(29.30)	Diagnostics
Celera Genomics Group	Nasdaq	683.00	105.10	139.70	13.40	Bioinformatics
Nektar Therapeutics Inc.	Nasdaq	447.80	94.80	157.80	(107.50)	Drug Delivery

Drug Agent Biotechs). Major services concern:

- collection, under request, of biochemical and molecular data both from public and private sources. Drug-oriented companies may use this service in order to run a first screening of possible targets in a research project;
- implementation of software systems to collect and manage data (eventually providing data mining tools);
- support for clinical tests (from a managerial and legal perspective) as well as lab activities, eventually carrying out lab tests directly. The latter are usually also called CROs (Contract Research Organisations);
- support for marketing activities.

Commodities Biotechs usually provide a "physical" output. They produce and sell consumable products (e.g. reagents and basic chemicals) for the drug research and development process. In most cases, products of Commodities Biotechs have an high degree of specialisation (i.e. they specifically suit the characteristics required by the buyer). However, these products actually give a marginal contribution to the overall value of the final products.

Finally, if all the typologies of companies (both pharmas and biotechs) are considered, it is possible to draw a complete map of the positioning of such actors on the value chain of the bio-pharmaceutical sector (Fig. 2.11).

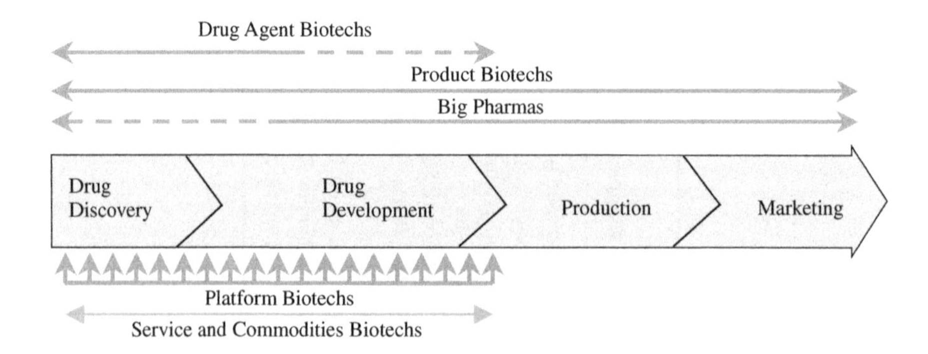

Fig. 2.11: The positioning of bio-pharmaceutical companies on the value chain.

2.4 The Industry Structure: A Geographical Analysis

To complete the overview of the bio-pharmaceutical industry, a geographical analysis looking at 2002 data has been conducted. Country data (Table 2.7) clearly identify Europe and US as leading countries in the development of the biotech sector. Despite huge investments in recent years, indeed, Asia still plays a marginal industrial role in the sector.

The US is commonly acknowledged as the "homeland" of biotechnology from at least three points of view: (i) historical (Genentech, the first biotech company, was founded in 1976 in San Francisco); (ii) financial (US have the greatest amount of investments in biotechnology); and (iii) normative (since 1980 a clear legal framework for biotech applications has been adopted).

The tables below (Tables 2.8–2.10) show some figures concerning the European and US biotech industries. A comparison between average data is also provided.

Europe had 1,878 biotech companies (with $ 13.7 billion of total revenues) in 2002, whereas US biotechs numbered 1,466 (with $ 28.5 billion of total revenues). Despite a lower number of companies, however, the value created by US biotechs is more than three times the value of European companies.

Most US companies are Product or Drug Agent Biotechs, whereas lower value-added models still prevail in Europe. This represents particularly a relevant threat for the future development of the biotech sector in Europe.

A deeper insight on Europe (Fig. 2.12) reveals analogous differences among the European countries, and particularly supports the evidence that

Table 2.7 Global biotechnology industry data.

Area	Number of companies	Employees
Europe	1,878 (102 listed)	87,182
USA	1,466 (318 listed)	142,900
Canada	493	7,785
Asia	601	9,764

Table 2.8 The US biotechnology industry.

	2002	2001	2000	2002/2001
Turnover (mln. $)	32,500	28,500	25,700	14.0%
Revenues (mln. $)	n.d.	20,700	18,700	n.d.
R&D expenses (mln. $)	n.d.	15,700	13,750	n.d.
Net losses (mln. $)	n.d.	6,900	6,000	n.d.
Funding (mln. $)	10,448	7,900	32,700	32.3%
Number of companies	1,466	1,457	1,374	0.6%
Employees	n.d.	191,000	176,000	n.d.

Table 2.9 The European biotechnology industry.

	2002	2001	2000	2002/2001
Turnover (mln. $)	13,787	13,730	9,870	0.4%
R&D expenses (mln. $)	7,524	7,480	5,520	0.6%
Net losses (mln. $)	1,550	1,520	1,810	2.0%
Number of companies	1,878	1,879	1,734	−0.1%
Employees	n.d.	87,182	67,445	n.d.

Table 2.10 Comparison between US and Europe (average data).

	2002		2001		2000	
	Europe	US	Europe	US	Europe	US
Turnover (mln. $)	7.3	22.2	7.3	19.6	5.7	18.7
R&D expenses (mln. $)	4.0	n.d.	4.0	10.8	3.2	10.0
Net losses (mln. $)	0.8	n.d.	0.8	4.7	1.0	4.4
Employees	46	132	46	131	39	128

a clear distinction may occur between the leadership in value creation and in the number of companies.

The German leadership by number of biotech companies (nearly 380) is mostly due to Platform, Service and Commodities Biotechs, whereas

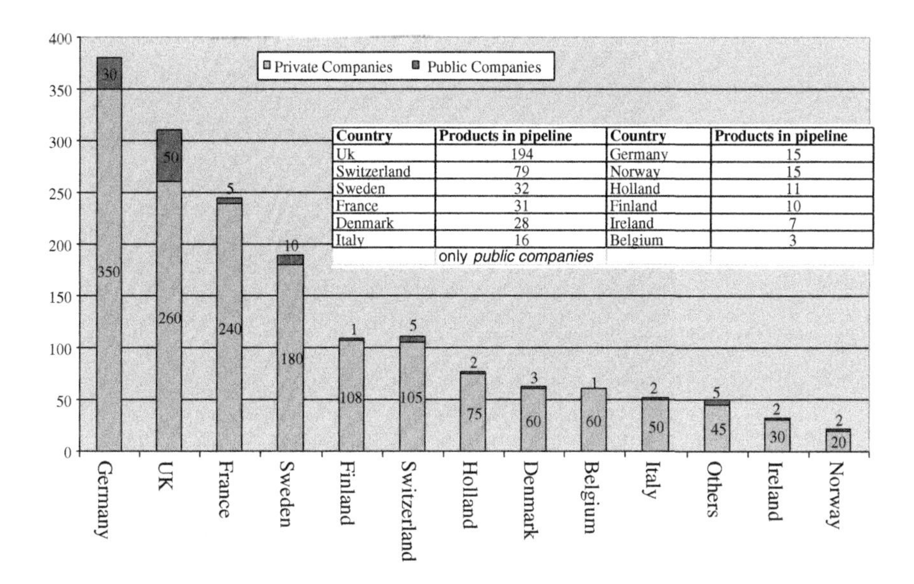

Country	Products in pipeline	Country	Products in pipeline
Uk	194	Germany	15
Switzerland	79	Norway	15
Sweden	32	Holland	11
France	31	Finland	10
Denmark	28	Ireland	7
Italy	16	Belgium	3

only *public companies*

Fig. 2.12: European biotech companies distribution.

UK companies (less than 320) are consistently focused on the new drug development and even production. With 154 potential new drugs in their pipelines, indeed, UK retains the leadership in added-value creation as compared to their German counterparts, which account for only 11 new drugs in pipeline.

3 The Cluster of Cambridge

3.1 History of the Cluster

The industrial biotechnology cluster within Cambridge emerged in the early '80s in a high-tech environment created by existing electronics and computing industries. Initial companies were founded within the Cambridge Science Park (owned by Trinity College, University of Cambridge), which itself was built to attract computing companies. A national strategy paper was published in the mid '70s by the UK Government with the intention of making universities more proactive in industry, and this resulted in the creation of initial science park buildings by Trinity College. These were, as mentioned, not built specifically for biotechnology companies and, moreover, the College did not build them with a view to spinning out its own science. Rather, they were built with the College acting as landlord only. Now, the Park is dominated by biotech companies and viewed primarily as a biotech location. The availability of scientific premises was supported by a change in attitude from some major investors within the Cambridge area. Barclays Bank, one of the largest banks in Britain started investing in more high-tech industry and venture capitalists followed suit. The number of biotechs grew steadily until the mid-'90s, when a global explosion of investment in high-tech industries accelerated company creation at a sustained rate until the stock market decline in 2001/2.

The success of the Cambridge Science Park spawned the development of additional research sites, including Granta Park, Melbourne Science Park and, more recently, Cambridge Research Park and Cambourne Business

Park. Already possessing a world-leading biotech research profile through organisations such as the University of Cambridge, the Institute of Biotechnology and the Babraham Institute, the cluster received a significant boost through the location of the major European effort for the Human Genome Project at the Sanger centre, located within the Wellcome Trust Genome Campus in addition to EMBL-EBI (European Bioinformatics Institute).

Development of the cluster continues today, with a continuing need for laboratory space driving development of science parks, and the University of Cambridge adopting a more proactive approach to commercial application of academic research.

As the cluster has evolved over the last two decades, the critical mass of industrial biotechnology has attracted an equal weight of technical and business service providers, creating a cluster rich in academic and commercial science, well served by local support providers.

With over 160 biotechnology companies and a greater number of service providers, the Cambridge cluster has achieved a mass that yields some shelter from global storms and has created a fully served community attracting investors' interests from across the globe.

The cluster is perhaps unique in Europe as no one person or organisation has consciously played a significant role in the cluster's creation or development. Several factors combined almost spontaneously to create an environment conducive to life science company start up. A number of biotechnology entrepreneurs were focused in Cambridge at the time and the mixture of increased funding, availability of premises, a high-tech atmosphere and altered attitudes to risk resulted in pioneer companies such as Celsis. Many of the figures involved in the cluster launch are still in Cambridge today, founding more companies and helping to perpetuate the biotech cluster. Any new biotech will have either, as a founder or on its board, a figure that has featured in commercial biotechnology for at least 10 years.

At the same time, since the cluster inception, a number of organisations help to create and support companies in a variety of fashions, even though none could be credited with a pivotal supporting role. For example, the Babraham Bioincubator provides small laboratories and office units on flexible leases and many of Cambridge's newest companies find their feet in this convenient location. The Bioincubator provides little or no subsidised services as is typical of incubators supported by Government funding and, indeed, was created many years after the cluster developed. ERBI (East Region Biotechnology Initiative) acts as a networking and cluster promotion organisation, and indirectly contributes to cluster cohesion and identity through its networking meetings and annual conference. However, as with Babraham, ERBI was created many years after the Cambridge cluster was recognised and provides a service supported on purely commercial foundations.

3.2 Major Actors

The analysis takes in account: (i) the DBFs; (ii) the industrial and research environment; and (iii) the financial environment.

3.2.1 Dedicated Biotech Firms

Overview

Cambridge has currently over 160 DBFs and has built a strong cluster profile as a centre for early stage companies with high growth rate and innovative technologies. Moreover, within the region, there are some examples of large biotech companies. Traditionally, the large company category would be exclusively pharmaceutical firms. However, biotech companies such as Celltech and Cambridge Antibody Technology have to be classified as large companies through an employee number greater than 250. These companies have a very different profile from pharmaceutical companies, operating on biotechnology models rather than the classical pharma structure, e.g. Cambridge Antibody Technology has no large-scale manufacturing and its first product for registration is being handled through a partner.

For the rest of the chapter, the small DBFs, in particular, will be analysed. All general information on Cambridge's DBFs are shown in Table 3.1. Figure 3.1 illustrates the speed at which the Cambridge cluster developed. As indicated by the number of pre-1984 companies, cluster

Table 3.1 General information on the Cambridge cluster.

Number of DBFs present in the cluster	162
of which local DBF	144
of which foreign DBF	18
Number of public firms	14 BioFocus, Bioglan Pharma, Cambridge Life Sciences, Celsis, Theratase, GeneMedix, Nexan Group, Pharmagene, Phytopharm, Acambis, Alizyme, Weston Medical Group, Xenova, Cytomyx.
Number of profitable firms	2 — BioFocus; Acambis.
Total employees of the DBFs of the cluster (2002)	9,600

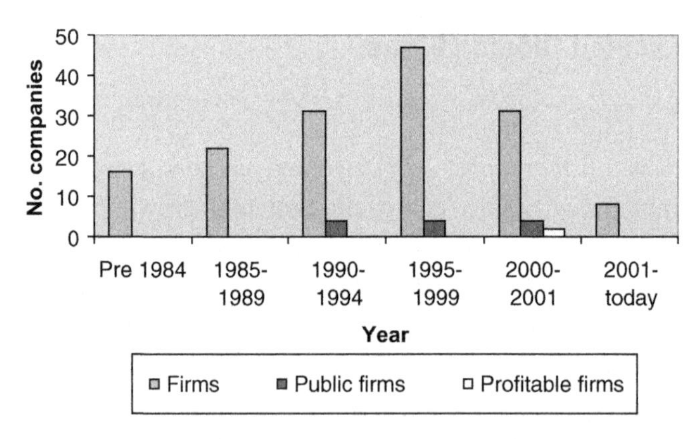

Fig. 3.1: Evolution of the number of firms, of profitable firms and of public firms in the Cambridge cluster.

origins were in the early '80s. However, the most rapid period of growth is 1995–1999, as a result of two main factors: (i) commercial awareness of biotechnology boomed during this period and it suddenly became feasible to start a biotech company either from academic start-ups or big pharma origins; and (ii) a rapidly growing global economy fuelling large venture capitalists investment in a broad range of companies.

The speed of start up was maintained through the millennium until the funding window closed in late 2000 and the global economy started to falter. Companies are still being started now but investment is far more cautious and this is reflected in a drop in the rate of company growth. Within the DBF community, a good number of firms is publicly quoted on a stock exchange (an alternative stock exchange was commonly selected for initial floatation before the London Stock Exchange relaxed its rules). Following a buoyant global economy and high company valuation, the bulk of flotation took place within the 1995–1999 boom period and has dropped dramatically since then as the market has slumped, taking company values with it. Further flotation is not anticipated until 2004 at the earliest and this is dependant on a market upswing.

The low number of profitable firms within the Cambridge DBF cluster is indicative of the cluster structure, a concentration of early stage, highly innovative healthcare companies. The goal on bringing a company to product (and profit) stage is some way in the future for many of these companies, and some will never reach it, having been acquired or run of money. The Cambridge cluster, indeed, has long been associated with therapeutic applications (Fig. 3.2).

There is a strong human therapeutic focus that stems back from the first products produced within the cluster in the '80s. This is despite a traditionally strong agricultural research base within the region surrounding Cambridge. This has been driven by the attractiveness of technologies to investors, with spending on human health continuing to accelerate, unlike that in agricultural applications.

With regard to the product/platform orientation of the Cambridge cluster, company focus has been strongly defined by levels of investment available and the global economic environment. In its infancy, before biotechnology was viewed with enthusiasm by investors, output from Cambridge was very product-based. Companies had to follow the fastest route to financial return on research, this being therapeutic products.

As the world awoke to the potential of such technologies and investment in the UK took off in the mid-'90s, the Cambridge cluster witnessed

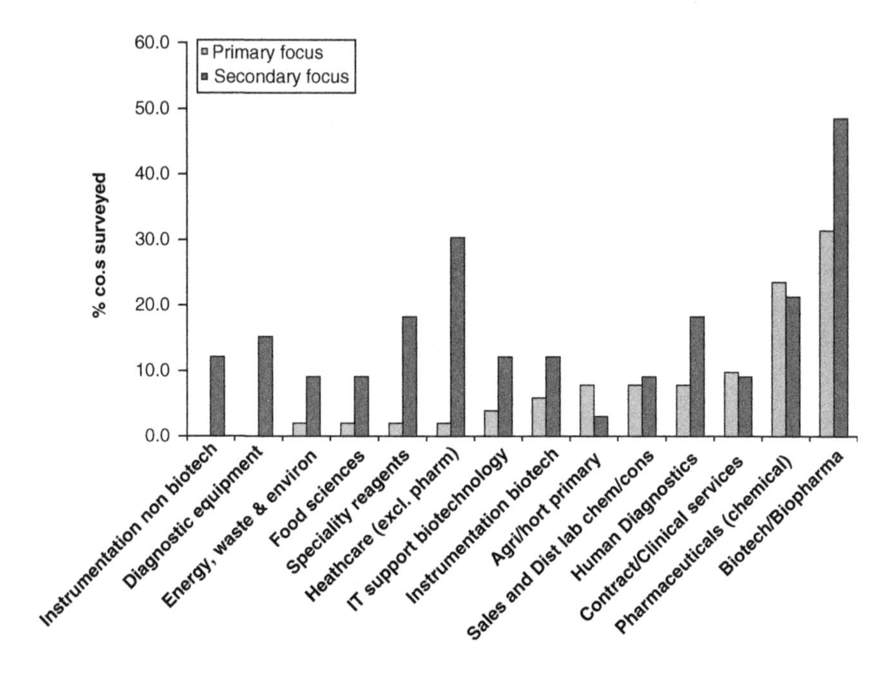

Fig. 3.2: Primary and secondary focus of companies in the Cambridge cluster.

the birth of a significant number of platform technology companies, organisations that, rather than produce products directly, developed technologies that enabled other companies to produce more effective products faster.

As the funding window closed in 2001, the Cambridge cluster has seen a shift back towards the product-based economy as companies strive to create value for investors through the development of therapies. Many companies now combine an innovative platform technology, which is promoted to larger biotechs and pharmaceuticals, with their own in-house discovery and development programmes.

Research and employment profile

Evolution of R&D expenses cannot be detailed for all DBFs within the Cambridge cluster. However, Table 3.2 gives an indication of changes in R&D spending over a three year period (2001–2003), highlighting that an increasing amount of resources is invested in R&D projects.

Table 3.2 Evolution of R&D expenses in the Cambridge cluster.

	2001	**2002**	**2003**
Mean annual expenditure in R&D activities (€ million per company)	5.04	6.21	6.40

The past five years has seen rapid expansion of biotechnology companies with up to 10% increase in company size per year, far exceeding the UK average. However, this is altered drastically since 2001 as companies look to minimising costs and operating for as long as possible on a finite research budget. Overall, company expansion has halted within the cluster. This halt is composed of a combination of some companies reducing numbers, others still expanding and the majority operating a recruitment freeze.

As a result of these trends, the Cambridge Cluster is currently typified by the most common company size as being between 11 and 20 employees (Fig. 3.3). This is a strong indication of the overall structure of the cluster; large numbers of small to medium innovative drug discovery and development companies looking to out-license technologies to larger companies well equipped to manufacture and market products.

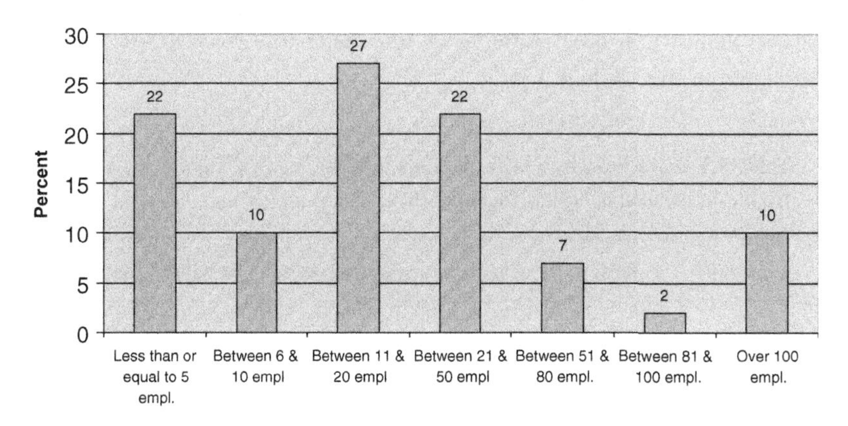

Fig. 3.3: Percentage distribution of firms by number of employees in 2002 in the Cambridge cluster.

The high number of companies with less than 5 employees indicates the continuing drive to start up new companies, particularly as the University of Cambridge becomes commercially more proactive within the cluster.

Process of foundation

The majority of biotech SMEs within the Cambridge cluster were born within it rather than becoming established from external sites. However, several external companies moved to the area, even if virtually none have moved their entire operation to the region from elsewhere, rather they have established additional research laboratories within the cluster. This has been done in two ways:

- *Opening of a completely new operation within the cluster.* This is a high risk strategy for a company as it will involve significant expenditure for an expansion of existing capabilities. Companies that do this have a strong draw to the region. Cyclacel, for example, is a biotechnology company based in Dundee, Scotland that opened a Cambridge arm after one of its founders moved to Cambridge to take up a position within the University;
- *Becoming established through the merger with or acquisition of an existing Cambridge company.* This is the primary way of becoming established within Cambridge as it has the additional benefit of acquiring new technologies/products in an already functional organisation. Examples of companies that have undertaken this route to a Cambridge presence are listed in the Table 3.3.

Table 3.3 Biotechnology SMEs that have moved into the Cambridge cluster from external national or international sites.

Company	Country	Local company acquired/merged with	Year in
ArQule	US	Camitro	2000
Xenova	US	Cantab	1989
Accelrys	US	Cambridge Molecular Design	
BioFocus	UK	Cambridge Drug Discovery	2001

Motivation to move to the Cambridge cluster comes from a number of sources: (i) critical mass reached by cluster with some of the strongest companies in Europe based there; (ii) strength of scientific research across the healthcare sector; (iii) base for entry into rest of Europe; (iv) significant pool of scientists and managers; (v) excellent profile for investors; (vi) good quality of life, attractive to employees; and (vii) excellent spread of service companies, both technical and business.

Concerning the local DBFs, it is possible to say that the Cambridge cluster has reached a level of maturity that makes definition of the origin of its companies very hard. The pool of people involved in founding new companies is reasonably small and while almost all are originally from a scientific industrial or academic background, they may have not been formally associated with those organisations for many years.

The Cambridge cluster has a strong history of biotech start-ups by people directly from an academic background but with no University links with the new company. This reflects the approach taken by the University of Cambridge towards use of research for commercial purposes. For many years, while the University maintained a policy of allowing commercialisation, this policy was not proactively encouraged. The presence of large pharmas within the region also provided fertile ground for industrial start ups and spin offs. The consolidation through merger of the pharma industry over the last decade has released non-core technologies and skilled managers/scientists to start their own companies. An excellent example of such a spin off is Adprotech, formed in 1997 by managers from SmithKline Beecham (now GlaxoSmithKline) and utilising intellectual property deemed non-core. Management buy-outs as a founding point for biotechs is rare in the Cambridge cluster. External parent companies do not form a significant proportion of the biotech community and closures undertaken by pharma parents are usually too large to allow a management buy out.

3.2.2 Industrial and Research Environment

As noted, the cluster of Cambridge is characterised by the presence of important industrial actors, both "traditional" pharmas and large biotech companies, as summarised in Table 3.4.

Table 3.4 Large companies in the Cambridge cluster.

Company	Employees in the cluster	Year of establishment in the cluster	Status in Cambridge cluster
Bayer plc	n.a.	n.a.	One of multiple sites
Bio Products Laboratory	500	1978	Single site
Cambridge Antibody Technology	260	1996	Single site
Celltech Group	140	1980	One of multiple sites
DuPont	10		One of multiple sites
GlaxoSmithKline	2,200	1715	One of multiple sites
Merck Sharpe & Dohme	1,700 in UK	1953	One of multiple sites
Millennium Pharmaceuticals	130	1997	One of multiple sites
Napp Pharmaceuticals	700	1984	One of multiple sites
Monsanto	n.a.	n.a.	One of multiple sites
Sigma-Genosis	70	1992	One of multiple sites
Syngenta	n.a	2001	One of multiple sites
Unilever Research	950	1930	One of multiple sites
Unipath	450	1984	One of multiple sites
Incyte Genomics		1998	One of multiple sites

Subsidiaries of large companies within the region have tended to take the form of acquisitions to add a particular technology to a global portfolio, e.g. Monsanto acquired PBL and Millennium acquired Cambridge Combinatorial. It is increasingly the case that local subsidiaries are more vulnerable when conditions are less favourable for the parent company.

Many of the large companies undertake R&D within the region, although for historical rather than cluster-linked reasons. Manufacturing also has a presence within the region, although it is unlikely that any large company would chose a Cambridge production site in the future, given the expense of the region compared to manufacturing centres such as Malaysia.

Besides a strong industrial base, the research environment in the area of Cambridge represents a key driver for the development of the cluster itself.

The Cambridge cluster is synonymous with the University of Cambridge and it is closely linked with the biotechnology cluster. It could not be described as having been instrumental in the formation of the cluster from a commercial angle; the University has been slow to realise the commercial potential of its research and only now do we see start up companies in significant numbers from the university. However, the University has been critical in the provision of world leading research and researchers since long before the cluster emerged. Trinity College, within the University has the strongest commercial links with the biotechnology cluster as it developed and owns the Cambridge Science Park. As already described, it developed the park to cater for high-tech companies in the computing sector before realising the biotechnology potential of the site. Moreover, the academic prowess of the University was certainly a key factor when the decision was made to site the Human Genome Project in Cambridge and it continues to attract world class research investment (e.g. Microsoft). The number of companies linked to the University through founders and spin-offs demonstrates the rich vein of intellectual property and skilled researchers that are based within the academic side of the cluster (Table 3.5).

With regard to the biotechnology capacity of the university, its scale is demonstrated by the fact that it has almost 40 departments linked to biotechnology, each with a clearly defined research area. In recent years, moreover, the University has become more active in encouraging researchers to commercialise the results of their work, thus reinforcing its impact on the cluster development. On one side, the Cambridge University developed specialised departments (Entrepreneurship Centre, Technology Transfer Office, Corporate Liaison Office) to promote the commercialisation of research. On the other side, incentives were set to encourage the commercialisation, e.g: (i) CUE (Cambridge University Entrepreneurs) Business Plan and Business Creation competitions, where teams can win €1,600 or €80,000 for business plans for new businesses respectively; and (ii) Challenge Fund which provides financial assistance for researchers planning to commercialise their work.

Other than University, the Cambridge cluster has many research organisations associated with it, some of which have a high profile within the cluster (Table 3.6).

Table 3.5 Biotech spin-offs of the University of Cambridge.

Akubio (2001)
Sensitive acoustic detection technology for rapid, label-free screening of drugs and diagnosis of viral and bacterial infections.

Astex Technology (1999)
A structure-based drug discovery company using high throughput X-ray crystallography technology for the rapid identification of novel drug candidates.

Avidis (2000)
Recombinant Protein Expression, Folding and Design for Therapy, Structural Analysis and Drug Discovery.

Biotica Technology (1997)
Combinatorial biosynthesis of therapeutic polyketides.

Biotransforms (1997, previously known as Pollution Technologies)
Bacterial DNA cloning and evolution directed strain selection technologies with applications in chemical bio-sensing, bio-remediation and chemical bio-transformations.

Cambridge Bioclinical (1997)
Research and development for pharmaceutical products for hair and skin therapy.

Cambridge Biotechnology (2001)
An innovative drug and drug target discovery company.

Cambridge Combinatorial (1996, sold first to Oxford Molecular Group, which was then bought out by Millennium Pharmaceuticals)
Combinatorial Chemistry.

Cambridge Drug Discovery/Cambridge Genetics (1997, now part of BioFocus plc)
Provided technologies in medicinal chemistry and biological screening with specialized expertise in kinase, GPCR and ion-channel targets.

Cambridge Microbial Technologies (1999)
Developed a series of original proprietary technologies for optimising protein expression systems.

Chroma Therapeutics (2001)
Pharmaceutical R&D into regulation of gene expression.

Clinical & Biomedical Computing (1999)
Informatics.

De Novo Pharmaceuticals (1999)
Structural and ligand based drug design with chemical synthesis and biological testing. Compounds from completed in-house research are also available as out-licensing opportunities.

Table 3.5 (*Continued*)

KuDOS (1997)
To discover and develop innovative products that modulate human DNA repair to treat human disease, particularly cancer.

Metris Therapeutics (1996)
Developing innovative treatments for gynaecological conditions.

Microbial Technics (1994)
R&D and Intellectual Property licensing company. Original proprietary technologies for optimizing protein expression systems (lactic acid bacteria).

Paradigm Therapeutics (1998)
Novel drug targets.

Sense Proteomics (1998, previously known as Sense Therapeutics)
Developing and using the next generation proteomics technology — functional protein microarrays — for successful drug discovery.

Smart Bead Technology (2000)
SBT develop massively parallel testing technologies, e.g. bead based bioassays, for the genomics, drug discovery and development, and diagnostics markets.

Solexa (1998)
DNA sequencing technologies and development of methods for chemical analysis at molecular level.

Among the listed organisations, there is the only incubator in the area, the Babraham Bioincubator. It was launched by the Babraham Institute and currently hosts 19 start up and early stage biotechnology companies.

3.2.3 Financial Environment

Almost without exception, start-up money is provided by venture capital, both national and international. Business Angels are beginning to take an interest in biotechnology funding, particularly with the opening of Library House (a facility designed to act as a focal point for investors) but their contribution is still low compared to venture capital. Commercial banks play little or no role in starting up biotechnology companies because of their high-risk nature and this shows no sign of altering.

Table 3.6 Research centres linked with the Cambridge cluster.

Organisation	Research activity	Commercial activities
Babraham Institute	Signalling, developmental genetics, molecular immunology, neurobiology	Babraham Bioscience Technologies Ltd. Acts to patent and license the Institute's research, manages commercial contracts and the Babraham Bioincubator.
Cranfield Biotechnology Centre	Medical diagnostics, environmental diagnostics, food biotechnology, combinatortial and computational chemistry, microbiology, sensors	Cranfield undertakes contract research and commercialises its own technologies.
European Molecular Biology Laboratory — European Bioinformatics Institute	Research and services in bioinformatics. The Institute manages databases of biological data including nucleic acid, protein sequences and macromolecular structures generated by HGMP	The EBI provides training to industry. It also provides microarrays of publicly available DNA and technology to facilitate gene clustering and sequencing.
Institute of Biotechnology	Part of Cambridge University, research incl. molecular biology, metabolic engineering, protein engineering, microbiology, biotransformations, enzyme technology, biosensors, combinatorial chemistry	The Institute offers entrepreneurial training for scientists. Research groups also have a well established history of commercialising research. It has originated a number of biotech companies.
Institute of Food Research	Food safety, diet and health, food materials and ingredients	The IFR Enterprise Unit coordinates all technology transfer and commercial work
John Innes Centre	Biological chemistry, cell and developmental biology, crop genetics, disease and stress biology, molecular biology	The technology transfer company, Plant Bioscience Ltd. handles all patenting and licensing of Institute research

Table 3.6 (*Continued*)

Organisation	Research activity	Commercial activities
Laboratory of Molecular Biology (LMB)	Structural studies, protein and nucleic acid chemistry, cell biology and neurobiology	It has originated a number of biotechnology companies
The Wellcome Trust Sanger Institute	Genome sequencing as part of HGMP	Publicly owned data

Finally, public funding is not available in Cambridge and this represents quite an exception in the European biotech clusters. The UK Government offers no investment to UK biotechnology firms. Small-scale research programmes are available (SMART, LINK Programmes) but these are for individual research projects and are relatively low cost programmes, thus being not commonly used by biotechnology companies of any size. "Soft" money is available elsewhere in the UK, in Objective One areas where the Regional Development Agency has deemed it suitable but Cambridge does not qualify for Objective One.

The strong presence of venture capitalists investing in biotech may be summarised by Tables 3.7 and 3.8.

The collapse of the stock market (occurring after the listing of the 14 biotech companies of the cluster — the last in 2000) has had both positive and negative effects on VCs. The negative aspect is the loss of investment funding as companies fail or share prices drop. The positive aspect is that VCs can now demand a large portion of the company in return for modest investment. Once the market improves, the venture capitalists stand to obtain a significant benefit. The overall result is that the flow of investments in the biotech sector is still growing.

3.3 Context Factors

The biotechnology industry, despite a high profile and favourable Government opinion, is not targeted directly by large-scale Government initiatives. Most of the effort is from the central Government through the Department

Table 3.7 National venture capital funding in the Cambridge cluster.

Venture capital firm	Cluster companies (€ invested where known)
Merlin Venture	Cyclacel (€6.4 million over 2 rounds), Adprotech Ltd. (€9.6 million over 2 rounds), Amedis Pharmaceuticals Ltd (€6 million over 2 rounds), De Novo Pharmaceuticals (€7.3 million in 1 round), Cambridge Biotechnology (part of €6 million 1st round funding)
Close Finsbury Asset Management (invests in plcs)	Celltech, Alizyme, Cambridge Antibody Technology
Prelude	Adprotech, CeNeS, De Novo Pharmaceuticals, Acambis
Avlar	Amedis Pharmaceuticals, Amura Ltd, De Novo Pharmaceuticals Ltd, Paradigm Therapeutics Ltd, Cambridge Biotechnology Ltd, Proteom Ltd, Therasci Ltd
Abingworth	Akubio,Astex, Solexa, Lorantis, Hexagen (now Incyte), Cantab (now Xenova)
Advent Venture Partners	Ribotargets, KuDOS Pharmaceuticals
Quester	Ribotargets, Avidex, Cyclacel, De Novo, Lorantis
Technomark Medical Ventures	Biotica
Providence Investment Company	Biotica
Generics Asset Management Limited	Biotica
Northern Venture Management	BioFocus, Alizyme, Cyclacel
Cambridge Gateway Fund	Cambridge Biotechnology, De Novo Pharmaceuticals

of Trade and Industry (DTI) and within that, the Office of Science and Technology (OST). Regional initiatives are not dictated by central Government although the funding is often provided centrally. It is a matter for the regional body to decide whether any initiatives are launched that target the biotechnology industry.

Table 3.8 International venture capital funding in the Cambridge cluster.

Venture capital firm	Cluster companies
3i	Cambridge Antibody Technology, Celltech, Weston Medical, Adprotech, Ribotargets
EuclidSR Partners	KuDOS Pharmaceuticals, Cambridge Antibody Technology, Millennium Pharmaceuticals
Johnson & Johnson Development Corporation	KuDOS Pharmaceuticals
LSP-Life Sciences Partners	KuDOS Pharmaceuticals
BankInvest Biomedical Venture	KuDOS Pharmaceuticals, Smartbead, Cyclacel, Biotica
Apax	Ribotargets, Celltech, Ionix Pharmaceuticals, Sense Proteomics, Xenova
Kargoe LLC	Ribotargets
NIB Capital	Ribotargets
Rendex	Ribotargets
JP Morgan	Ribotargets, Celltech
OrbiMed	Ribotargets
Biotechnology Value Fund	Biotica
Nordic Biotech	Biotica

It is the general perception from the biotechnology industry, that public bodies have little or no impact on the industry. The primary concern of companies is to obtain funding and, as noted, public funds are not available within the Cambridge cluster. No particular strategy has been followed by regional or national public bodies to develop the cluster and it has developed purely on commercial lines. Naturally, European funding is open to Cambridge biotech companies and this has been sought with varying degrees of success.

Concerning the legal framework, few specific laws regarding biotechnology research exist in England. The most important laws have reference to use of animals and ethical issues such as stem cell research and cloning, while data protection is covered primarily by European laws. The major biotech-related laws in England are:

- Animals (Scientific Procedures) Act 1986;
- Human Fertilisation and Embryology Act 1990 (the 1990 Act).

The second, in particular, is unique to English laws as the UK has a more flexible approach to stem cell research using embryos than most other European countries, granting licenses through the Human Fertilisation and Embryology Authority (HFEA) to use embryos less than 14 days old.

This approach of the English government is the result of the general good level of interest in stem cell research and other recent technologies with strongly perceived health benefits. Most biotechnology products linked to healthcare are perceived favourably, with the exception of some religious groups and pro-life groups (for the case of embryonic stem cells and gene therapy).

The role of the media has been critical in establishing such acceptance as the public is strongly influenced by media opinion. This can be demonstrated by the media approach to GM crops which was, and still is, extremely hostile, resulting in a general backlash against the technology and a moratorium on GM crop trials announced by the UK Government. The Government is influenced by public opinion, often more strongly so than its advising scientists, and this demonstrates the power of the media in the UK political process. The current Government has, on the whole, been very supportive of biotechnology, which has protected the industry from more negative media coverage and key public figures such as Lord Sainsbury and Lord Winston have been instrumental in creating a positive scientific profile in the public eye.

3.4 Conclusions

Cambridge has currently over 160 DBFs and has built a strong cluster profile as a centre for early stage companies with high growth rate and innovative technologies. Moreover, within the region, there are some examples of large biotech companies: Celltech and Cambridge Antibody Technology, for example, have an employee number greater than 250. The cluster of Cambridge emerged in the early '80s as one of the first European area with a significant concentration of biotech firms. The most rapid period of growth of the cluster is 1995–1999, as a result of two main factors: (i) commercial awareness of biotechnology boomed during this period and it suddenly became feasible to start a biotech company either from academic start ups or big pharma origins; and (ii) a rapidly growing global economy fuelling large venture capitalists investment in a broad range of companies. In the meantime, a good number of firms gained the access to the capital market. The global economic downturn in late 2000 reduced the speed of start up and of IPOs, thus enhancing the first signs of consolidation within the cluster.

Concerning the business models, the DBFs within the cluster generally present a strong focus in human therapeutics, even if some "deviations" from this focus had been historically experimented because of the levels of investment available and the global economic environment. In the mid-'90s, the Cambridge cluster witnessed the birth of a significant number of platform technology companies (more near-term oriented). Currently, it is interesting to observe the predominance of "mixed" business models, where companies combine an innovative platform technology with drug discovery and development programmes.

The favourable background for the birth and development of the Cambridge cluster can be recognised in its strong scientific base. Even if the University of Cambridge has indeed only recently became a major player in the direct creation of new biotech companies (academic spin-offs), the excellence of research of its more than 40 departments linked to biotechnology is acknowledged at international level since the '80s. The contemporary presence of important industrial actors, both "traditional" pharmas and large biotech companies, contributed to increase the growth rate, but in a secondary role.

The analysis of the history reveals the peculiar spontaneous and commercial nature of the cluster. The birth and development of the area of Cambridge come from the untidy set of initiatives of actors of different nature (DBFs, large companies, universities, ...) without a strong commitment of public actors, which did not act under a common strategic scheme. Even if in a lack of a common scheme and of the direct public intervention, it is possible to recognised in the life cycle of the cluster (that is currently in its maturity) some driving forces that triggered the cluster birth and then its further development.

Considering separately each stage, in the birth phase the following driving forces can be highlighted:

- *availability of seed and venture capital.* Despite the lack of effective public funding programmes in the area, there was a great number of venture capitalists (both local or international, with businesses in the area) investing in biotech and of business angels and other small investors focused on the seed stage financing;
- *presence of mechanisms to attract key scientific people.* This contributes in creating a leading-edge scientific base. The University of Cambridge strongly invested in the specialisation of its biotech-based departments, reaching the excellence in different scientific and technological fields;
- *diffusion of entrepreneurial culture among scientists.* Even if, as noted, the direct creation of spin-offs from the University of Cambridge came later, since from the beginning the commercial "taste" of its researchers led to the birth of many independent start-ups, exploiting the results of successful researches.

In the development phase, and still nowadays, the key factors of the cluster are:

- *networking culture*, that is the establishment of close relationships within universities and research centres and between these ones and existing companies in the geographical area of the cluster;
- *international promotion of the cluster*, through a centralised strategy that comprises the creation of "cluster representatives". Such strategy may help in the attraction of new sites form external companies, offering them a clear motivation to move to the cluster of Cambridge.

In order to perform these activities, in the development phase, many agencies were created, such as the BIA (Bioindustry association) the ERBI (East Region Biotech Initiative), the EEDA (East Anglia Development

Agency). These organisations that now act as central actors were created when the biotech industry was already well established and "conscious" of being in a cluster. Therefore their birth occurred significantly later than the one of their counterpart in the clusters where these organisations are a direct consequence of the public intervention.

Therefore, this aspect also differentiates the Cambridge case from the other European clusters analysed. In the Cambridge cluster, neither strong public intervention nor shared actions among actors were necessary to the birth and development of the cluster. The context presented from the beginning all the critical resources needed for cluster success.

4 The Cluster of Heidelberg

4.1 History of the Cluster

The biotech cluster in the Rhine-Neckar Triangle (also "*Bioregion Rhine Neckar Triangle*") is located in an area between Neustadt (in the south), Heidelberg, Mannheim and Ludwigshafen (in the centre), Darmstadt (in the north) and Kaiserslautern (in the west), comprising a region with a radius of about 40 km. The centre of this cluster is Heidelberg, with more than 30 biotech companies in the "Technologiepark Heidelberg" and an excellent scientific background from universities and national and international research institutes. In the area there is one of the biggest concentration in the world of leading-edge research bodies in the field of molecular biology: the University of Heidelberg, the European Laboratory for Molecular Biology, the German Cancer Research Centre, and the Max Planck Institutes for Medical Research and for Cell Biology. However, research activities within the cluster concern a broader range of scientific and technological areas. A strong research base is focused on genomics, proteomics, bioinformatics, neurobiology, immunology, and virology. Moreover, the geographic area of the cluster includes industrial and research sites of multinational pharmaceutical companies (like Merck, Roche Diagnostics, Abbott, etc.) as well as of medium-size companies (Becton Dickinson, Hyland Immuno, Lion Bioscience, etc.) strongly involved in the pharmaceutical biotechnology. Finally, more than 80 dedicated biotech firms (most of which had been founded in the last 5 years) complete the biotech-related industrial base of the cluster.

The history of the development of the cluster was strongly affected by two main elements: (i) the creation of the Technology Park; and (ii) the BioRegio contest.

The Heidelberg Technology Park was founded in 1985, with the commitment of the major public actors of the region (the City of Heidelberg and the Chamber of Industry an Commerce Rhine-Neckar), which took an equity position. The "concept" of the public actors in the founding process was to support the commercial exploitation of the research results, leveraging the existent excellence in the most innovative branches of life sciences. Expected outcomes were the creation of new jobs, the application of technical and scientific know-how, and the development of the economic and competitive position of the region. The Heidelberg Technology Park started offering laboratories, office spaces, daily supports and services as well as opportunities to exchange ideas and experiences to start-ups and established companies, with a strong focus on life sciences. Initially, 11 companies were located in the park. In 1998, the available space for the Heidelberg Technology Park grew from 12,000 m^2 to more than 16,000 m^2. Another expansion, in spring 2003, added nearly 32,000 m^2: so there are currently near 50,000 m^2 available for companies to start and to expand. A second site, called the "Production Park", is close to the main station area and dedicated to growing companies. Additional opportunities for the expansion of the biotech sector in the area of the cluster are the TZ Ludwigshafen and the space for biotech companies in Mannheim as well as other laboratory buildings spread in the region.

The BioRegio contest actually boosted the cluster development but, moreover, had a strong impact on the whole biotech sector in Germany. Compared to other countries, especially the US and the UK, biotechnology had a slow start in risk averse Germany. Falling behind in a so-called "generic" high-tech industry was a matter of serious concern for German policymakers in the early '90s. Therefore, the BioRegio contest, launched by the federal government in 1995 with a €75 million budget, was designed to transform a "dormant" sector into one intended to be globally competitive by stimulating biotech firm start-ups, the growth of existing companies and the provision of venture capital. Two elements have been particularly important: (i) the elements of evaluation in order to determine the winners; and (ii) the conditions in which financings had to be assigned.

Concerning the evaluation, all regions wishing to participate in the contest had to give a presentation of their respective strengths in biotech as well as proposals for future development of biotechnology in the region. An independent jury was installed by the Federal Research Ministry to find the three best organised regions with the most promising development concepts on the basis of the following criteria: (i) number and scale of existing companies oriented towards biotechnology in the region; (ii) number, profile and productivity of biotech research facilities and universities in the region; (iii) interaction (networking) of different branches of biotech research in the region; (iv) supporting service facilities (patent office, information networks, consulting); (v) strategies to convert biotechnology know-how into new products, processes and services; (vi) regional concept to help the start-up of biotechnology-based companies; (vii) provision of resources through banks and public equity to finance biotechnology companies; (viii) cooperation among regional biotech research institutes and clinical hospitals in the region, and (ix) local authorities approval practice with regard to new biotech facilities and field experiments.

Governmental funds were mostly dedicated to funding new companies. Funds were available for DBFs' requested financing only if they were also able to collect at least the same amount of money from private investors. Many regions went further from this offering, creating dedicated biotech funds. Due to this mechanism, which lowered the risk of the financing (many local actors gave an additional insurance that if biotech start-up failed they were paying part of the private investment), venture capitalists "jumped" into the biotech context.

In some regions the local or state governments coordinated the regions' activities, in other cases it was industry or research institutions themselves. In all regions enterprises, research institutes and government officials cooperated very closely. Seventeen BioRegions formed to participate in the contest. Some of them are single cities (and their hinterland) such as Freiburg, Jena or Regensburg, while others are networks of neighbouring cities such as Braunschweig-Göttingen-Hannover or

Heidelberg-Mannheim-Ludwigshafen, or they cover whole federal states such as Berlin-Brandenburg.

The three regions selected by the jury as winner regions were Munich, Rhineland, including the cities of Cologne, Aachen, Düsseldorf and Wuppertal, and the Rhine-Neckar Triangle with Heidelberg, Mannheim and Ludwigshafen. It was pointed out that these regions all have a comprehensive scientific basis in modern biotech research, substantial entrepreneurial activity in the field of biotechnology and a promising regional development concept for biotech industry. The East German region of Jena received a "special vote" for its "especially positive new-orientation" in the field of biotechnology after re-unification.

Public funds amounting to DM 150 million (nearly €75 million) are reserved for the three winners in the BioRegio contest. Moreover, the winning regions had priority in the appropriation of funds from the "Biotechnology 2000" program of the Federal Research Ministry for a time span of five years. The latter advantage seems to be the more important one since the total amount of public biotech funding in Germany (about €750 million from 1997 to 2001) is about ten times higher than the direct BioRegio award and the jury's judgement on the regions' capability and concepts is of crucial importance for the spatial distribution of funds from the larger budget.

The most important advantages of the BioRegio contest, acknowledged by the participants themselves, appear to be the enhancement of communication and cooperation among the regional key actors, the establishment of a regional environment favourable to the innovation, the furthering of research cooperation within the BioRegions and the stimulation of interregional competition for technology. Moreover, there was a "self-consciousness effect": the regional actors have become aware of their region potential and the social acceptance of biotech within the regions has improved.

The Rhine-Neckar Triangle won the competition on November 20, 1996, thus starting a new growth phase: within the BioRegio framework, approximately 130 applications have been reviewed between 1997 and the end of 2001, of which 38 projects with a total volume of €56 million were selected to receive the BioRegio funding. Most of the supported projects were in the fields of functional genomics, pharmacogenomics, proteomics, bioinformatics and related technologies. The Rhine-Neckar Triangle "BioRegion" concept foresees three main activities: (i) assessing commercial potential within the scientific institutes; (ii) establishing a business-run organisation to exploit this potential and help with project financing; and (iii) taking measures to encourage business skills among scientists and acceptance of biotechnology in the general public.

A crucial factor in winning the BioRegio competition was the idea for a Biotechnology Centre in Heidelberg as the focal point of efforts to channel academic ideas into successful products and services. Three key organisations strongly contribute to the establishment of this idea: (i) Heidelberg Technology Park, mentioned above; (ii) BioRegion Rhine-Neckar-Dreieck eV; and (iii) Heidelberg Innovation.

The main objective of the BioRegion Rhine-Neckar-Dreieck eV during the BioRegio competition was to evaluate the business plans of candidate biotech companies. Currently, the BioRegion Rhine-Neckar-Dreieck eV is better involved in the organisation of events to improve political and social favourable conditions for the biotech sector, as well as supporting initiatives for companies looking for available industrial spaces as well as public grants.

Finally, Heidelberg Innovation, founded with the objective of supporting technology transfer mechanisms and evaluating business plans, recently re-focused their activities in managing seed and venture-like capital funds dedicated to the biotech industry within the cluster.

Moreover, technology transfer offices were created in the major universities and research centres within the cluster in order to favour the exploitation of research results and still play a key role in the management of IPRs and in the licensing procedures in which the research bodies are involved.

4.2 Major Actors

The analysis takes in account: (i) the DBFs; (ii) the industrial and research environment; and (iii) the financial environment.

4.2.1 Dedicated Biotech Firms

Overview

At the beginning of the BioRegio contest, in 1996, there were in the area 31 DBFs and 4 sites from big pharmaceutical companies (BASF, Merck,

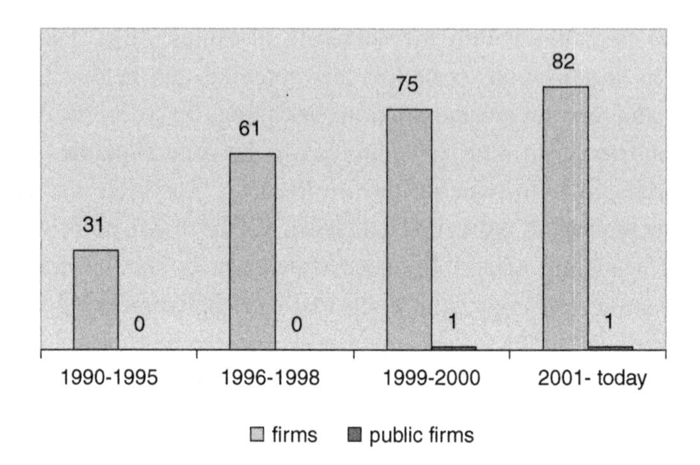

Fig. 4.1: Evolution of the number of firms and of public firms in the Heidelberg cluster.

Abbott and Roche). Thanks to the participation to the competition, 51 new companies (with more than 1,200 new jobs) focused on biotechnology were founded, thus enabling the strong establishment of the biotech industry in the Rhine-Neckar Triangle and, ultimately, the creation of the Heidelberg cluster.

The analysis of the evolution of the number of firms in the cluster (Fig. 4.1) strongly underlines the effects of the BioRegio contest in the cluster development. The number of foundations per year experienced a huge growth after the winning announcement of the BioRegio competition: changing from only 4 new DBFs in 1997 to 10 (+250%) in 1998 and 16 (+60%) in 1999. In recent years, the negative economic conjuncture and the first signs of consolidation in the cluster reduced the number of new companies (14 in 2000 and 7 in 2001).

The main characteristics of the DBFs in the cluster are shown in Table 4.1.

It is of particular interest to notice the followings:

- nearly 94% of companies are generated by local entrepreneurs. The large predominance of DBFs created by local entrepreneurs is a direct consequence of the strong commitment of public actors in the region and of the evaluation criteria set up in the BioRegio contest;

Table 4.1 General information on the Heidelberg cluster.

Number of DBFs present in the cluster	82
of which local DBF	77
of which foreign DBF	5
Number of public firms	1 Lion Bioscience
Number of profitable firms	11 Progen; BioReliance; Biopharm; Becton Dickinson; Immundiagnostik; Orpegen Pharma; Alfatec-Pharma; IBL Umwelt- und Biotechnik; BioCat; Promega; The Binding Site
Total employees of the DBFs of the cluster (2001)	1,700

- nearly 13% of companies gain profits. Despite a high percentage of profitable firms, however, there are problems in accessing the capital market;
- only 1 company (Lion Bioscience) is public. The number of public companies, therefore, is significantly lower than the one experimented in the other comparable clusters (e.g. in the cluster of Cambridge, the percentage of public firms is near to 10%). Among the causes, the fields of application in which the DBFs operate can be highlighted.

Table 4.2 shows that the majority of companies are platform companies (i.e. they concern the technological development of physical and "logical" devices to be used in the biotech research activities), thus enabling a low-risk profile and profits in the near-term, but at the same time encompassing low growth perspective in long-term horizon. Moreover, 10% of companies carry out their activities in the fields of veterinary and environmental biotechnology, which currently play a marginal role (both for the number and diffusion of possible applications). Capital markets, instead, show a huge preference for healthcare-based businesses, which are widely acknowledged with high performance in the long-term.

Table 4.2 Main fields of application of DBFs in the Heidelberg cluster.

Main fields of application	% of firms (*)
Healthcare	39
Agro-food	0
Nutraceuticals	0
Veterinary	3
Environment	7
Others (technology platform, bioinformatics, etc.)	51
Total	100%

(*) data refer to the 51 companies founded in the period 1995–2001.

The relative "youth" of the large majority of DBFs in the cluster and their start-up phase in most cases "forced" by the BioRegio competition provide another reason to the mentioned difficulties to reach the capital markets. The evaluation criteria set up by the Federal Research Ministry, particularly the one concerning the regional concept to help the start-up of biotechnology-based companies, forced local public actors to stimulate scientists and researchers to write business plans and to start-up new ventures, even if they are in the early stages of their research.

Turnover and employment profile

Turnover are very low (Table 4.3). Despite the fact that 30% of companies refused to provide their data, the figure that emerges from the analysis is quite surprising: 60% of companies reported in their 2001 annual reports turnovers worth less than €1 million.

Among the causes of such low turnovers, it is interesting to notice the effects of the above mentioned strong stimuli from the BioRegio competition in the creation of new ventures. The large majority of these new companies rely for their revenues only on partnerships or licensing agreements and presents products in the early stages of development. For example, all the DBFs in the cluster directly involved in the drug development are currently performing, with their products, the preclinical phase, thus resulting in a

Table 4.3 Turnover of DBFs in the Heidelberg cluster.

Turnover 2001	% of firms
Below 1€ million	60
Between 1€ mil. and 2€ mil.	
Between 2€ mil. and 3€ mil.	
Between 3€ mil. and 4€ mil.	
Between 4€ mil. and 5€ mil.	
Beyond 5€ mil.	10
n.a.	30

Table 4.4 Number of patents in the last three years in the Heidelberg cluster.

	1999	2000	2001
Patent applications	n.a.	21	420
Registered patents	5	6	73

weak stream of revenues (eventually from royalties). Despite the low profile of returns, however, DBFs in the cluster have hardly started to invest in R&D activities (even if, as mentioned, in particular on technologies rather than drugs), with 73 registered patents in 2001 out of 420 applications, more than ten times the ones registered in 1999 (Table 4.4). Moreover, near 40% of companies had registered more than 10 patents in the period 1999–2001.

Concerning the employment profile, the number of employees grew continuously from 1997 to 2001 and currently there are about 1,700 employees in the DBFs of the cluster. Table 4.5 presents the data of the companies founded between 1996 and 2001, showing an average number of employees that grew from 17 per company in 1999 to 24 per company in 2001.

Process of foundation

The analysis of the process of foundation of DBFs in the cluster helps understand the current situation of the biotech industry in Heidelberg.

Table 4.5 Number of employees (1999–2001) in the Heidelberg cluster.

Firm	1999	2000	2001
A3D	2	4	5
Abeta	—	1	7
Acconovis	—	—	4
Affimed Therapeutics	2	10	11
Anadys Pharmaceuticals	—	—	14
Apogenics	2	2	2
ATEC	8	18	17
Axaron	55	67	86
BioCat	—	4	4
BioGenerix	—	7	18
bioLeads	25	38	44
BioReliance	18	17	20
Biotrin	4	4	4
BMI Biomedical Informatics	—	6	6
Bts BioTech Trade & Service	—	—	2
Cellzome	—	30	100
Cenix Bioscience	1	12	0
CLONTECH	17	17	17
Cytonet	—	25	38
Dr. Gottschall Instruction	9	14	14
Febit	30	43	72
GeneArtists	—	—	3
Generatio	—	10	10
Graffinity Pharmaceutical Design	20	29	53
Heidelberg Engineering	—	—	26
Heidelberg Pharma	3	18	23
The	10	28	42
Hybaid	30	20	20
ISIS optronics	5	6	6
LION bioscience	132	265	320
LYNX Therapeutics	6	17	17
Medical Communications	9	12	10
Microcuff	—	—	6
MMI	14	14	16
MRC-Systems	12	20	25
MTM Laboratories	8	18	18
Neuroscience & Pain	—	2	4

Table 4.5 (*Continued*)

Firm	1999	2000	2001
Peptide Special Laboratories	—	1	1
phase IT	—	—	12
Phytoplan	7	8	8
Pointer Pharmaceuticals	—	—	1
Promega	18	29	30
PromoCell	4	4	4
ROOTec	4	4	4
STM	—	—	2
Symbiosis	5	5	4
Therascope	—	17	20
UV-Systems	15	17	17
VISCUM	—	5	5
Wise Gene Products Innovation	2	5	5
20/10 Perfect Vision	6	10	10
Total	506	882	1,207

Table 4.6 Year of foundation of DBFs in the Heidelberg cluster.

Year of foundation	Number	Cumulative percentage
Before 1996	35	41
1996–1999	30	76
1999–2001	21	100
Total	86	

Tables 4.6 and 4.7 present the year and the process of foundation of the DBFs in the cluster.

It is interesting not only to notice the large predominance of start-ups, but also the huge presence of spin-offs from the research environment (universities and research centres) that are a direct consequence of the BioRegio competition. Despite the presence of large sites from big pharmas, instead,

Table 4.7 Process of foundation of DBFs in the Heidelberg cluster.

Process of foundation	% of the total
Start-up	52.8
Industrial spin-off	13.7
Academic spin-off	8
Scientific spin-off	23.5
Joint Venture	2

- Start-up: a new company which has not any formal relation with existing industrial or scientific entities;
- Industrial spin-off: a firm that has some formal relations with an industrial actor (parent company);
- Academic spin-off: a firm that has some formal relations with a university;
- Scientific spin-off: a firm that has some formal relations with a previous research centre;
- Joint Venture: a firm formed by the formal collaboration between two other actors;

the industrial environment played a marginal role in the generation of the DBFs. Moreover, industrial spin-offs are the earliest founded companies and actually receive a little (or no) "spin" from the national biotech competition.

4.2.2 Industrial and Research Environment

The industrial environment related to biotechnology in the Heidelberg cluster is constituted, as previously mentioned, by the sites of 4 large pharmaceutical companies (BASF, Merck, Abbott and Roche Diagnostic). Particularly, the industrial sites established in Heidelberg focus respectively on the followings: BASF in agricultural and industrial biotechnology, Merck in recombinant proteins, Abbott in the treatments for rheumatoid arthritis, and Roche in the active substances.

The profile of the research environment in the area is significantly higher, with a strong presence of universities, research centres and a large science park (Heidelberg Technology Park).

The situation of universities is presented in Table 4.8. The presence of nearly 1,000 people in 2001 involved in biotechnology is "exploited"

Table 4.8 Academic environment in the Heidelberg cluster.

	1999	2000	2001
Number of universities with biotech department	3	3	3
Total number of researcher in the biotech sector	ca. 40	ca. 50	ca. 60
Total number of student in biotech courses	ca. 800	ca. 850	ca. 900
Total number of graduated people in biotech	ca. 150	ca. 170	ca. 200
Number of technology transfer offices	1	1	1

for the development of the cluster itself through direct policies to foster entrepreneurship implemented by the University of Mannheim, the University of Heidelberg and the Academy for Advanced Education of the Universities of Heidelberg and Mannheim.

The cluster area includes some of the most renown research institutions at international level in the fields of molecular biology and molecular medicine. The European Molecular Biology Laboratory (EMBL), the German Cancer Research Centre (DKFZ) and the Centre of Molecular Biology of the University of Heidelberg (ZMBH), as well as the Max-Planck Institute for Medical Research (MPI), the research institutes and university hospitals of Heidelberg and Mannheim all contribute to the excellent scientific base in biotechnology.

The main characteristics of some of these centres are briefly presented here. EMBL currently employs 750 researchers in the fields of molecular biology, biophysics, gene expression and bioinformatics. DKFZ, founded in 1964 as a non-profit organisation and supra-regional research centres by the Land of Baden-Wuttemberg, operates, with 650 researchers, primarily in oncology diagnostics and experimental therapy. MPI is an independent, non-profit organisation, established in 1948. Research highlights over the years have included major contributions to the analysis of the first metabolic pathway-glycolysis, a broad range of breakthroughs involving muscle biochemistry and structure, elucidation of the bioenergetic role of ATP, and unraveling of the mechanisms of cell regulation and DNA replication. Recently, MPI pioneered the development of synchrotron radiation for biological research.

Finally, the Heidelberg Technology Park, already described in the history of the cluster and currently hosting 38 companies, represents the "engine" of the cluster development, providing a full range of equipped spaces as well as services for the incubation and the development of biotech companies.

4.2.3 Financial Environment

The main sources of funding for the biotech companies in the Heidelberg cluster are the governmental investments that came in 1996, totalling €25 million, as a consequence of the winning of the BioRegio contest by the Rhine-Neckar Triangle. Accordingly to the competition rules, the governmental funds directly contribute to the start-up of 38 projects out of 120 presented applications together with the contribution of private investors: the total amount was nearly €54 million. In the meantime, since 1996, many other smaller competitions and federal and regional funding initiatives have been developed, providing an effective and efficient source of seed capital for the establishment of new companies.

The support of venture capitals, both local (Heidelberg Innovation and ETF) and foreign (3i, Atlas Venture, Burrill&Co, Sofinnova, etc.), strongly influenced the growth of the companies within the cluster. In particular, venture capitals provided the start-ups with their second and third round financing, thus making them able to survive after the initial "public-driven" momentum (Table 4.9).

The current negative financial context, with the IPO window firmly closed, represents an actual threat for the future development of the cluster: the time horizon becoming longer and longer and the reduced returns' expectations may lead to a crisis of the funding system.

4.3 Context Factors

The German government, leveraging a favourable social environment, since 1990 (five years before the start of the BioRegio competition) settled a well defined and generally favourable legal framework concerning biotechnology (particularly "red biotechnology"). Among the others, two laws are

Table 4.9 Venture capital funding in the Heidelberg cluster.

	1999	2000	2001
Number of local VC that invest in biotech firms	1	1	2
Number of national VC that invest in biotech firms of the cluster	4	5	8
Number of foreign VC that invest in biotech firms of the cluster	n.a.	3	5
Average investment of local VC in local biotech firms	0.5 M€	0.5–0.75 M€	more than 1 M€
Average investment of foreign/national VC in local biotech	0	n.a.	2 M€

particularly important in defining "the boundaries" of the biotech sector: the Genetic Engineering Act and the Embryo Protection Law.

The Genetic Engineering Act (Gentechnik-Gesetz, GenTG) of July 1, 1990 can be better referred to as a set of laws aimed at clearly defining the legal standard authorisation of laboratories and production facilities, as well as field trials with genetically modified organisms. The German Genetic Engineering Act incorporated the key elements of three directives of the European Union: (i) the "Anwendungs- oder System-Richtlinie (90/219)", a regulation, which concerns the use of genetically modified organisms (GMOs) in industrial and research facilities; (ii) the "Freisetzungs-Richtlinie (90/220)", which governs the deliberate release of GMOs into the environment; and (iii) "The Schutz der Arbeitnehmer gegen Gefaehrdungen durch biologische Arbeitsstoffe bei der Arbeit (90/679)", which regulates workers safety. This set of initiative solved many local problems relating new manufacture facilities authorizations, which strongly affected the development of German biotech context in 1980s. In addition to this, the law has a federal effect, making the legal context more uniform, even if some characteristics remained at local level, such as many elements of the approval process. The German government introduced the GenTG legislation to achieve a balance between industrial and environmental concerns. Initially, the Act was viewed by the industry as a step in the right direction: certain products such as Interleukin-2 and Erythropoietin (EPO) that are manufactured using innocuous microorganisms or cell culture technology could be relatively easily produced. However, it soon became clear that the bureaucratic, extensive and often too stringent regulations embodied in the Act would continue to stifle German biotech industry, affecting its ability to compete internationally. As a reaction to heavy lobbying from the biotech industry and the continued exodus of research and production facilities abroad, the German government finally agreed to amend the Act in 1993 to reduce restrictions and ease the approval processes. Two major amendments to the GenTG loosen and thereby simplify some of the restrictions, easing the way for a more rapid development of the commercial biotech sector in Germany, thus paving the way for the 1995 BioRegio competition.

The development of the biotech sector allowed by the Genetic Engineering Act, however, did not divert the government from question regarding the "ethical boundaries" of research. In 1990, the German government approved the Embryo Protection Law, which totally bans biotech activities using embryos. Stem cell regulations in Germany have long been criticized for being too stringent. The production of new human embryonic stem cell lines is illegal in Germany. The experimental use of human embryonic stem cells is restricted to cell lines generated before January 1, 2002, from surplus embryos resulting from infertility treatment. Research involving such existing stem cell lines requires the approval of a national review committee. It is important to notice, however, that this law, currently under revision, does not have strong effects due to the analysed major focus of German biotech firms on platform technology or service development.

Finally, clear procedures, both at national and regional level, are settled concerning IP rights. At national level, the "Arbeitnehmer Erfindergesetz" regulates the procedure for inventions in companies, universities and other institutions. In Baden-Württemberg (where Heidelberg is located), the rights of university inventions belong to the inventors. They may choose to carry out the whole procedure of patent application and eventually surveillance autonomously, or they may allow the universities to forward the invention to the Technologie- und Lizenzbüro (Technology Transfer Office), which evaluates the invention, applies for the patent and negotiates for the licensing agreement. In the latter case, TLB, university and inventor share the earned licensing fees.

4.4 Conclusions

The biotech cluster in the Rhine-Neckar Triangle is centred in Heidelberg and comprises the area among the cities of Neustadt (in the south), Mannheim and Ludwigshafen (in the centre), Darmstadt (in the north) and Kaiserslautern (in the west). The cluster has currently a radius of about 40 km and 82 dedicated biotech firms operate in it.

The current strong biotechnological industrial base, however, was mainly created in the last five years, given the fact that the Rhine-Neckar Triangle won in 1996 the BioRegio competition, thus starting a new phase of growth. At the beginning of the competition, indeed, there were in the area only 31 DBFs. The number of foundations per year experimented a huge growth after the winning.

Nearly 53% of DBFs are start-ups, while 37% are spin-offs from the research environment (universities and research centres). Despite the presence of large sites from big pharmas (BAFS, Merck, Abbott and Roche Diagnostic), instead, the industrial environment played a marginal role in the generation of the DBFs.

Concerning the business models of the DBFs in the cluster is interesting to notice that the large majority of companies are platform companies, thus enabling a low risk profile and low growth perspective in long-term horizon, especially if compared to product based companies. Moreover, 10% of companies carry out their activities in the fields of veterinary and environmental biotechnology, which currently play a marginal role (both for the number and diffusion of possible applications) in the whole sector. As a consequence of this peculiar business orientation, despite a high percentage of profitable firms in the cluster, Heidelberg's DBFs face relevant "problems" in accessing the capital market (only one company, Lion Bioscience, reached the capital market).

The analysis of the history of the cluster shows that the excellence in research was a key original factor. In the area of Heidelberg there is one of the biggest concentration in the world of leading-edge research bodies in the field of molecular biology: the University of Heidelberg, the European Laboratory for Molecular Biology, the German Cancer Research Centre, and the Max Planck Institutes for Medical Research and for Cell Biology. Moreover, research activities within the cluster concern a broader range of scientific and technological areas (genomics, proteomics, bioinformatics, neurobiology, immunology, and virology). The excellence in life sciences research undoubtedly represented a favourable background for the development in the area of a cluster in the biotech sector. However, the history of the cluster reveals how a more complex set of driving forces effectively "triggered" the development process. Among the others, the followings can be highlighted:

- the public funding to the creation of new firms (through the Bioregio competition) directly aimed at stimulating private funding (particularly in the seed stage);
- the diffusion of entrepreneurial culture among scientists and academics, leveraging the existence of a strong research environment;
- the presence of dedicated infrastructures (the Heidelberg Technology Park), offering hosting services for the new biotech companies; and
- the presence of a clear and well defined legal framework (particularly the Genetic Engineering Act), facilitating the exploitation of research results in the life science area.

It is interesting to notice that all the mentioned factors are strongly related to the intervention of public actors. The Heidelberg Technology Park was founded in 1985, with the commitment of the major public actors of the region (the City of Heidelberg and the Chamber of Industry an Commerce Rhine-Neckar). The Bioregio competition, settled in 1995 by the German central government, represented for the area the opportunity to leverage both its excellence in life sciences research and its tradition in sustaining the technology transfer trough dedicated infrastructures. The availability of dedicated public funds, conditioned by the availability of at least the same amount of private funds from venture capitalists or business angels, led to the establishment of a "virtuous circle" where scientists are "forced" to become entrepreneurs and where the presence of the Heidelberg Technology Park allowed them not to move from the area.

The Heidelberg Technology Park, therefore, played a pivotal role in the development of the cluster, becoming the central actor around which the cluster was created. The need to implement a common strategy and to present a unique report at regional level for participating at the BioRegio competition, indeed, forced in Heidelberg the aggregation of different interested actors around the existing structure of the Park. Obviously, the central actor "survived" at the end of the competition and, moreover, the range of its activities has been progressively enlarged, accordingly with the growth process of the clusters.

The case of Heidelberg represents a major example in Europe of how a biotech cluster can be created through a policy-driven approach (i.e. through the implementation by the public actors of direct initiatives, leveraging the existence of some favourable conditions, aimed at forcing the creation of new DBFs).

5 The Cluster of Aarhus

5.1 History of the Cluster

The city of Aarhus, located in the western region of Denmark known as East Jutland, has currently a total population of nearly 285,000 people and the county of Aarhus as a whole accounts for more than 630,000 inhabitants hereby becoming the second largest city in Denmark.

Ten years ago in Denmark there was only a handful of large Danish companies connected with the biotech and the pharmaceutical sectors (e.g. Novo Nordisk, H. Lundbeck and LEO Pharma). These companies were — and still are — mainly situated in the Copenhagen area. Currently, there are more than one hundred dedicated biotech firms (DBFs), mainly founded around the turn of the century and mainly established in the Oeresund/ Copenhagen region, which represent the greatest biotech industrial concentration in Denmark and it is also known as Medicon Valley. However, several of these companies (together with new local ventures) have also established facilities/offices elsewhere in Denmark, primarily close to national universities, with the Aarhus area being the most effective region so far.

The history of the cluster of Aahrus is very recent and it is undoubtedly embedded with the commitment of the academic environment. The Aarhus region is the major education and research centre of West Denmark with more than 30,000 students and 2,000 scientists at university level.

The strong position is a result of internationally acknowledged research derived from the University of Aarhus and the connected university hospitals (seven hospitals), covering all branches of healthcare. However, not only the university, but all the other major regional stakeholders, including the hospitals, the industrial and financial actors, the City of Aarhus and the County of Aarhus started few years ago to strongly support the sector, aimed at exploiting the Aarhus potential in the biotech sector. In particular, the foundations of East Jutland Innovation (a private company, formed in 1998 with public money, providing pre-seed and seed capital for more than 50% of the biotech companies in the cluster) and Incuba (a dedicated venture capital firm) had significantly strengthened the biotech context, recreating a dynamic of creation of new firms.

The result of the initiatives of these actors is the current industrial base, made up of 22 companies the great majority (95%) of which are academic spin-offs from the university and university hospitals (the Aarhus cluster is quite unique concerning this characteristic). The first biotech companies in the cluster were founded in the early nineties, while the large majority (19 out of 22) have been founded from 1998 until 2001, in correspondence with the global economic growth period. The companies are primarily established in Science Park Aarhus, which provides incubator services (laboratory and office facilities). Starting in 2001, however, a declining trend in the creation of biotech companies has been noted. Other than the global economic downturn, among the causes of such decline, the fact that Denmark implemented a new law regarding IPR at universities and public research institutes (rights belong directly to the inventor), which strongly affected the creation of new ventures, should be highlighted. In order to face these problems, and to support the development of the cluster of Aarhus, the BioMedico Forum was founded in 2001. This association, which currently has 365 members, aims at enhancing the development of the biotech sector in Aarhus by promoting regional, national and international networking. BioMedico Forum's members include the University of Aarhus, healthcare organisations, and most of the biotech companies located in the region. Moreover, a dedicated organisation, Investment Location Aarhus, acts as a "facilitator" for new, incoming companies that want to set up branches in the cluster.

5.2 Major Actors

The analysis takes in account: (i) the DBFs; (ii) the industrial and research environment; and (iii) the financial environment.

5.2.1 Dedicated Biotech Firms

Overview

Leveraging a good scientific base, the cluster of Aarhus, even if it is in an embryonic phase, includes few but promising high growth companies, 75% of which are located in the Science Park. Table 5.1 presents the main information about the DBFs in the cluster.

As noted, the industrial context appears with few small size actors without large companies. None of the DBFs has gone public, and it is doubtful if any of the companies will reach that stage within the next 5 years. However,

Table 5.1 General information on the Aarhus cluster.

Number of DBFs present in the cluster	22
of which local DBF	21
of which foreign DBF	1 (MWG Biotech)
Number of public firms	0
Number of profitable firms	6 DNA Technology, Loke Diagnostics, Pipeline Biotechnology, Profundis Biotechnology, Aros Applied Biotechnology, MWG Biotech (foreign)
Total turnover of the DBFs of the cluster	€1.90 million (2001)
Total employees of the DBFs of the cluster	81 (2001)
Average size in terms of turnover	€0.12 million
Average size in terms of employees	3.7

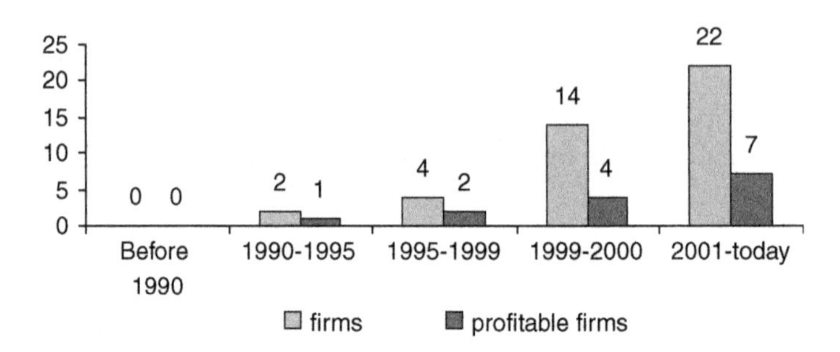

Fig. 5.1: Evolution of the number of firms and of profitable firms in the Aarhus cluster.

despite the limited amount of total turnover and the recent foundation of the large majority of DBFs, nearly 33% of them are already profitable.

Most of the companies were founded after 1999 (Fig. 5.1), mainly as a consequence of a change in the public funding mechanisms. In the mid-1998, it became possible to get loans from East Jutland Innovation on behalf of the Danish government for research-based entrepreneurs. The availability of pre-seed and seed capital managed at local level to foster the entrepreneurship among scientists within the cluster, triggered the creation of new ventures. Concerning the business models, it is interesting to notice that only 9 out of the 22 DBFs operate in the drug discovery process; 4 companies are specialised in diagnostic tools, while the others mainly provide support services (in animal testing and in bioinformatics). Obviously, given their near-term approach, the 7 profitable firms belong to the latter typology.

Turnover and employment profile

Total turnover appears very low, even if the analysis of the trend highlights a significant growth rate in last three years (Table 5.2). This remark confirms on the one hand the small size of DBFs in the cluster (only 4 companies presents a turnover between 1 and 2 million of Euro), and on the other hand, the high potential for the future development.

Also the number of employees (81 in 2001 in the whole cluster) reflects the small size of DBFs, with an average employment of nearly 4 people. Moreover, if the largest firm (DNA Technology) is not considered, the

Table 5.2 Turnover of DBFs in the Aarhus cluster.

1999		2000		2001	
Total	Average	Total	Average	Total	Average
€905, 405	€56, 588	€1, 040, 540	€65, 033	€1, 905, 405	€119, 088

average size of each company does even not reach 3 employees. In most cases, indeed, companies have only the inventor "employed" (Table 5.3).

Process of foundation

The cluster is dominated by academic spin-offs (Table 5.4). The University of Aarhus and the University Hospitals both have a long-standing tradition for research-collaboration with national as well as international pharmaceutical companies. This had historically led only in few cases to the direct commercialisation of research results by the academic environment, even if the Faculty of Life Science, the Faculty of Sciences and the University Hospitals have obtained a great amount of research grants and research contracts from pharmaceutical companies over the years. Only in the last decade of the century, and with a strong acceleration in the period 1998–2001, has the region developed a commercialisation strategy where research activities based in the region took form as academic spin-off companies.

The only "start-up" case in the cluster is the foreign company MWG Biotech. Its foundation process can be summarised as follows. In January 2000, MWG Biotech, headquartered in München (Germany), established its Scandinavian sales and research office in Aarhus. The office currently has 6 employees, comparing to nearly 400 employees for the whole company. MWG Biotech runs one of the worlds largest commercial sequencing facilities equipped with state-of-the-art technology and bioinformatics, thus offering a custom sequencing service and having expertise available for high throughput sequencing projects. Main motivations for moving into the cluster seem to be the potential cooperation with the academic and research environment in Aarhus.

Table 5.3 Number of employees (2001) in the Aarhus cluster.

Firm	Number of employees
DNA Technology	25
DNA Diagnostics	1
Loke Diagnostics	6
Borean Pharma	5
Recepticon	2
Plantic	1
ProteoTarget	1
Como Biotech	1
Action Pharma	1
Action Inflamation	1
ProteoPharma	1
CellCure	2
Cobento	1
Pipeline Biotech	8
ProSep	1
BioInformatics	1
Profundis Biotech	5
Aros Applied Biotechnology	3
Idac	1
MWG Biotech	6
Vivox	3
Stach International Institute	5
Total	81

The other DBFs, in most cases (80%) strongly supported by East Jutland Innovation, are still renting laboratory facilities at the University, as shown in Table 5.5, thus highlighting huge links with their parent insititutions.

Case studies

Borean Pharma

The technology currently exploited by Borean Pharma was initially developed in the Gene Expression laboratory at the University of Aarhus, as technology platform for a number of research projects relating to the structure

Table 5.4 Process of foundation of DBFs in the Aarhus cluster.

Process of foundation	Number of companies
Start-up	1
Industrial spin-off	0
Academic spin-off	21
Scientific spin-off	0
Joint Venture	0
Management buy-out	0

- Start-up: it is referred here as a firm that has not any formal relations with previous industrial or scientific entity.
- Industrial spin-off: it is referred here as a firm that has some formal relations with an industrial actor (parent company).
- Academic spin-off: it is referred here as a firm that has some formal relations with a university.
- Scientific spin-off: it is referred here as a firm that has some formal relations with a previous research center.
- Joint Venture: it is referred here as a firm formed by the formal collaboration between two other actors.
- Management buy-out: it is referred here as a firm formed on the acquisition by the management of a subsidiary of a pre-existenting firm.

and function of modular proteins. A first commercialisation of part of this technology occurred in 1993, in cooperation with the Denmark company Cheminova. The result of the deal was the establishment of a new company, Denzyme, which was three years later acquired by the UK biotech Cambridge Antibody Technology. In 2001, a further agreement between Cambridge Antibody Technology, the original founders of Cheminova and the pre-seed investor NOVI led to the creation of Borean Pharma. Among the terms of the agreement, there was the jointly endeavour to raise venture capital for the new company. As a result of this process, a first round financing of €10.7 million was closed in December of the same year 2001 (the first biotech/pharma venture investment carried out in the Aarhus region).

Borean Pharma expanded its total staff from 5 to 21 in 2002 and, in 2003 announced the purpose of acquiring another biotechnology company, which was also an academic spin-off of the University of Aarhus.

Table 5.5 Academic spin-offs in the Aarhus cluster.

Firm	Year	Resources given by university
DNA Technology	1992	None
Loke Diagnostics	1998	None
Como Biotech	1999	Using of academic laboratories
Action Pharma	1999	Using of academic laboratories
Pipeline Biotechnology	1999	None
ProteoPharma	2000	Using of academic laboratories
CellCure	2000	Using of academic laboratories
ProSep	2000	Using of academic laboratories
Aros Applied Biotechnology	2000	Using of academic laboratories
Borean Pharma	2001	Using of academic laboratories
Recepticon	2001	Using of academic laboratories
Plantic	2001	Using of academic laboratories
Action Inflamation	2001	Using of academic laboratories
Profundis	2001	Using of academic laboratories
Cobento	2002	Using of academic laboratories
BioInformatibiotechcs	2002	Using of academic laboratories

Pipeline Biotech

Pipeline Biotech is a pharmacological contract research organisation with the capacity to combine advanced molecular biology with pre-clinical research. Its primary objective is to feed pharmacological companies with the knowledge base acquired from focused research studies in the drug discovery and early development processes. Pipeline Biotech, founded in 1999, experienced a steady growth that enabled it to design and carry out special research programmes, particularly those concerning animal experiments. Pipeline Biotech is the first company in this field in Denmark to combine pre-clinical *in vivo* studies with molecular biology.

Cobento

Six scientists from the University of Aarhus and the Aarhus University Hospitals founded, in November 2001, Cobento Biotech, exploiting their strong background in vitamin B12 and related diseases, protein chemistry, molecular biology and plant transformation. Starting from these premises, Cobento currently produces recombinant proteins for a huge range of uses: for diagnostic devices as well as dietary supplements and medicine. The company leverages its know-how in the development and production of recombinant proteins from plants, being the only biotech company in Denmark growing transgenic plants.

DNA Technology

DNA Technology was founded in 1992 as the first biotechnology company in the Aarhus Science Park. The company currently is acknowledged as one of the leading manufacturers and suppliers of customer-defined oligonucleotides. More recently, the company started research programs aimed at evaluating different techniques involved in multianalyte sensing and measurements. A new architecture for array-based sensors has already begun to be applied, allowing a new approach to the company focus. The arrays manufacturing, indeed, better concerns the field of human diagnostic systems.

5.2.2 Industrial and Research Environment

The Aarhus region does not have any large company directly focused on biotechnology or large "traditional" pharmaceutical companies. However, within the cluster, there are some important firms in the food sector interested in these new technologies. Aarhus hosts the biggest dairy and meat company in Europe (Arla Foods), one of the top 5 world players in food ingredients (Danisco) and 2 international leading food technology companies (Aarhus United and BSP Pharma).

Arla Foods, headquartered in Aarhus, has more than 9,000 employees in Denmark and a turnover of nearly €5 million. Currently the company is cooperating with the University of Aarhus (Department of Structural and Molecular Biology) on separating and developing new proteins for food applications. Danisco, founded in 1989, is listed on the Copenhagen Stock Exchange, currently employees near 8,000 people and presents net sales for €2.39 billion. Its headquarter is in Copenhagen, but the largest R&D department is in Aarhus. Danisco develops and produces food ingredients, feed ingredients, sweeteners and sugar (Danisco's broad product portfolio includes emulsifiers, stabilisers, flavours, and sweeteners such as Xylitol

Table 5.6 Academic environment in the Aarhus cluster.

	1999	2000	2001
Total number of academic researcher in the biotech sector	624	627	620
Total number of student in biotech courses	3,505	3,729	4,120
Total number of graduated people in biotech	656	697	774
Number of technology transfer offices	0	2	2

and Fructose). Finally, Aarhus United, founded in 1871, has been among the world's leading producers of vegetable oils and speciality fats for the food industry and speciality fats and oleo chemicals for cosmetics and pharmaceutical industries for more than a century. Aarhus United currently has 483 employees and €1 million in 2001 turnover.

Concerning the research environment (Table 5.6), the cluster presents two main research institutions with a long tradition in life sciences: the University of Aarhus and the Aarhus University Hospitals. The University of Aarhus was founded in 1928 and basic research in biotechnology, molecular biology and medical research are carried out at the Faculty of Sciences and the Faculty of Health Sciences. The University of Aarhus has recently been acknowledged by the Danish National Research Foundation as one of the national centres in which "excellent quality of research" is carried out.

The University of Aarhus Hospitals is comprised of 3 hospitals in Aarhus, covering all areas of medical and surgical specialist care, along with the Psychiatric Hospital, which provide national and regional specialist services. Both institutions employee more than 600 researchers and biotech-based courses accounted in 2001 for 3,279 students.

As a result of the excellence in research, Jens Christian Skou was awarded in 1997 the Nobel Prize in Chemistry for his pioneering work in physiology (iontransporting enzyme) and Lars Fugger was awarded in 2002 the prestigious Descartes Prize for a project tackling multiple schlerosis.

Besides these two institutions, major departments of governmental research institutions are established in Aarhus, some of them conducting research in biotechnology. Among the others, the followings can be highlighted:

- the Danish Institute of Agricultural Sciences (DIAS), a research institution under the Ministry of Food, Agriculture and Fisheries. With nearly 1,100 employees, DIAS is one of the largest research institutions in Denmark embracing a broad range of agricultural areas through the integration of animal husbandry, plant production and biotechnical research. DIAS comprehends the Agro Business Park offering tenancies to research-based companies (start-ups as well as established companies) within the agricultural and food industry;
- the Department of Terrestrial Ecology, belonging to the National Environmental Research Institute. It conducts research and consultancy on different fields: (i) terrestrial ecotoxicology, including the effects of pesticides and other chemical substances; (ii) release of genetically modified plants, including the environmental effects of organic farming; and (iii) map of the effects of air pollution on sensitive ecosystems;
- the Department of Environmental Chemistry and Microbiology. It develops methods regarding the occurrence and environmental transformation of persistent organic pollutants as well as risk assessment methods for genetically modified organisms and microbial pesticides.

The strong commitment of all the industrial and academic actors, as well as of the local governement, to support the development of the cluster led in 2003 to the creation of a new dedicated science park (Biomedical Science Park). Together with the mentioned Aarhus Science Park; the Biomedical Science Park aims at offering incubator services as well as at exploiting synergies between clinical research and commercialisation. The Park, still under construction, is located closely to Skejby Hospital, one of Aarhus University Hospitals. The first phase of construction that had been completed in Skejby consists of 3,700 m^2 of space for offices, laboratories, animal and other shared facilities.

5.2.3 Financial Environment

The Danish government is currently investing in new biotech start-ups through the so-called Innovation Environment programme, providing pre-seed and seed capital through local dedicated agencies. In the area of Aarhus, the Innovation Environment programme led to the creation of East Jutland Innovation, particularly "triggering" the diffusion and the exploitation of the entrepreneurial culture in academic institutions within the cluster. After the foundation of East Jutland Innovation in 1998, 20 new companies emerged, 10 of which are directly funded by the agency. Moreover, governmental-funded post-doc programs and flexible laboratory rental in the Aarhus University play a fundamental role in the development of many small biotech academic spin-offs.

Besides the governmental interventions, other financial actors in the cluster of Aarhus operate on both sides: funding research projects within the university, and providing venture capital for the further development of biotech companies.

In particular, on the one side, the Aarhus University Research Foundation operates as a commercial organisation, supporting scientific research at the University of Aarhus. Each year the Foundation awards grants totalling nearly €5 million devoted to specific research projects as well as to infrastructural projects (e.g. the building of facilities in the science parks). On the other side, Incuba Venture and BankInvest Bio Venture, venture capital firms active in Aarhus, recently made some investments in cluster's companies for nearly €2 million.

5.3 Context Factors

The Danish action plan for biotechnology and ethics (BioTIK) forms the basis for a number of specific projects, the majority of which to be started in next years. The action plan can be divided into two areas of intervention: (i) international regulation and cooperation on biotechnology and ethics; and (ii) public debate and information on biotechnology and ethics in Denmark.

International regulation and cooperation

The Danish Centre for Ethics and Law published in 2002 a report that investigates existing international and European regulations on the application of biotechnology in the plants and foods area. The report also refers to the ongoing ethical debate in this area across Europe. As a result, recommendations are launched for an international convention on ethical principles for genetic engineering in the plants and foods area. At European level, the BioTIK aims at implementing a comprehensive and supportive strategy for biotechnology and ethics.

Public debate and information on biotechnology and ethics in Denmark

The other part of the BioTIK concerns projects on public debate and information in Denmark with the aim of implementing new methods to include citizens and consumers in the evaluation of the applications of biotechnology, thus strengthening the dialogue between scientists and society and between enterprises and consumers.

Besides this long-term strategic plan, the "Act on inventions at public research institutions" concerning the intellectual property rights policies is of particular relevance as it highlights the commitment of the Danish government. The purpose of this Act, applied to all inventions made after the 1st January 2000, is to ensure that research results produced by means of public funds shall be utilized for the Danish society through commercial exploitation. Under this Act, when universities or hospitals are commercialising an invention, the income is equally divided between the inventor, the inventor's department and the university/hospital. The equal distribution of income results in a strong incentive for both inventors and universities in exploiting business ideas.

5.4 Conclusions

The cluster of Aarhus is still in an embryonic phase. The cluster currently encompasses 22 DBFs, the large majority of which has less than 5 employees. The "small scale" of the biotech companies within the cluster is the result of two main factors: (i) most of companies (80%) were funded after 1999; and (ii) all the companies (except for MWG Biotech, which is the only foreign DBF established in the cluster) are academic spin-offs of the University of Aarhus and of the University Hospitals. In most cases DBFs have only the inventors employed.

Despite the recent formation, nearly 33% of DBFs are already profitable. A reason for this figure may be traced back to the analysis of business models. Only 9 out of the 22 DBFs operate in the drug discovery process, with 4 other companies specialising in diagnostic tools, while the others mainly provide support services (in animal testing and in bioinformatics). Obviously, given their near-term approach, the profitable firms belong to the latter typology.

The history of the cluster reveals how, given the lack of a biotech-related industrial base (except for some large companies operating in the agro-food field), it was the academic environment that provides a favourable background for the cluster birth. Indeed, the University of Aarhus and the University Hospitals both have a long-standing tradition for research-collaboration with national as well as international pharmaceutical companies.

Given the scientific base in the area, the main driving forces (i.e. the set of actions settled up in order to leverage the favourable background) that, among the others, "triggered" the creation of the cluster are the following:

- the public funding offering pre-seed and seed capital to biotech companies;
- the availability of specific infrastructures (science parks).

Concerning the first issue, the Danish central government decided to provide pre-seed and seed capital (i.e. to fund companies in their critical start-up phase) through the creation of dedicated agencies within the Innovation Environment programme. In the Aarhus area, the programme led in 1998 to the creation of East Jutland Innovation, which directly funded 10 out of 18 companies established in the cluster. Funds are available directly to inventors, thus supporting them in the decision to exploit their research results through the creation of new firms.

Besides the financial support in the early stages of development, another set of actions aimed at fostering the birth and development of a cluster in Aarhus concerns the availability of dedicated science parks (the Aarhus Science Park and, more recently, the Biomedical Science Park). Parks offer hosting services and shared facilities, thus helping new companies in reducing operation costs in their first stages.

The combined impact of both forces results in a highly favourable environment for the new biotech firms, making them able to survive the early-stage "cash burning" development phase. The birth of the cluster of Aarhus, finally, may be brought back to the intervention of a public actor focused on the direct financing of new ventures in the biotech sector.

6 The Cluster of Marseilles

6.1 History of the Cluster

The cluster of Marseilles is at a very embryonic stage of development. Indeed, although there is a number of dedicated companies and supporting entities in the field of biotechnology located in the geographical area around the city, the degree of interaction between these actors still remains rather low. Some of them may have close bilateral relations one another but there is no general and regular multi-lateral flow of cooperation among them.

Yet the actual configuration is worth describing because there is a strong impulse of the institutional actors within the growing cluster to stimulate and develop interactive functionalities and to bring the existing "groups" of actors to a further level of cluster maturity.

The first nucleus of the cluster has been the group of University Hospitals of Marseilles and the university teaching faculties related to them (Medicine, Pharmaceutics, Biology, etc).

Since 1968 life sciences have been developing in the strong academic environment of Marseilles and in particular in the Science Park of Luminy. The first visible impulse on biotechnology came from the CIML (Centre d'Immunologie de Marseilles-Luminy) when Immunotech (a DBF) was created in 1982 by some researchers of the CIML to design, develop and manufacture reagents based on monoclonal antibodies for research and diagnosis.

This was a major event in the French academic and research environment because this type of spin-off was (at that time) totally uncommon. The venture turned out to be a success and generated a lot of interest and attention both from the business and the academic side.

Despite the positive aspects of this first "model", there were no immediate followers; there were many "enthusiastic" observers but few of them tried to replicate the pioneers' experience.

Some years later, the existence of a very strong medical field with academic, research and clinical branches provided enough assets (among others) to initiate a first federative initiative: the "Cité de la Biotique", which was to be a real estate unit dedicated to biotechnology firms and activities. In a short time, however, because of different reasons (related to the socio-political conjuncture) the public support to this entity was disrupted, thus failing in the scope.

Immunotech, at the same time, continued its growth, seeking extended financial support. As a result, in 1997, the company became a subsidiary of the American Group Coulter, which in its turn merged in 1997 with the Beckman Group, becoming Beckman–Coulter Group. The initial founders of Immunotech then decided to launch a new start-up (Trophos), thus keeping their autonomy and capitalising their first success; some of the former managers and scientists followed them, creating two more DBF companies in 1999: Innate Pharma and Ipsogen.

Immunotech, even if in some way was "out of the group" because its own geographical gravity centre was displaced abroad, indirectly gave a new impulse to the cluster dynamic.

Meanwhile, the interest for biotechnology continued to be present and a more "active" approach in recent years allowed the creation of a number of dynamic start-ups, showing that the area has the potential to produce concrete ventures. Beside these companies that had a huge focus on biotechnology, clinical tests centres and shared animal facilities sprung up in the area near the city of Marseilles, supporting the development of the cluster itself.

Nowadays, biotechnology is widely diffused in the whole "Région PACA" (Région Provence-Alpes-Côte d'Azur), particularly around

6 poles:

- *Marseilles*: 2,352 researchers, engineers and technicians (59% of the total PACA resources). Moreover, recently some DBFs started their activities in the close proximity of Marseilles in Aubagne-Gemenos;
- *Avignon*: 700 researchers, engineers and technicians (18%), that is an important pole for agronomy;
- *Nice*: 535 researchers, engineers and technicians (13%);
- *Antibes*: 150 researchers, engineers and technicians (4%);
- *Toulon*: 150 researchers, engineers and technicians (4%);
- *Cadarache*: a research centre from the CEA (Commissariat à l'Energie Atomique), 100 researchers, engineers and technicians.

The current vitality of the sector in the area and the worldwide interest concerning the biotechnology market led some major local institutions to support a new initiative to federate local actors in the biotechnology field into a real dynamic cluster. The "Association du Grand Luminy", the CCIMP (Chamber of Commerce and Industry of Marseilles-Provence) and an economic "club" (the GT8 — Groupe de Travail 8) together with "Provence-Promotion" (Economic development departmental agency) joined a partnership initiative and sponsored a task team to organise meetings with the actors involved and to produce an opportunity study for a future biotechnology cluster. This task team has produced its report at the end of 2002.

The key characteristics of these organisations are here briefly resumed. The AGL (Association du Grand Luminy) was created in 1985 to support the development of the Luminy University Science Park. AGL has a board of directors including the representatives of the main research centres and academic/economic actors of the Luminy site. It is now one of the strongest actors in supporting the cluster and the editor of the feasibility study produced at the end of year 2002. AGL was created with the support of the CCIMP the first French "*enterprise centre*": 165 projects have been studied and 48 accepted in the enterprise centre leading to the creation of 17 firms in 5 years (over 100 jobs). In this context, a special attention has been paid to the promotion of biotechnology. Finally, Provence-Promotion, is

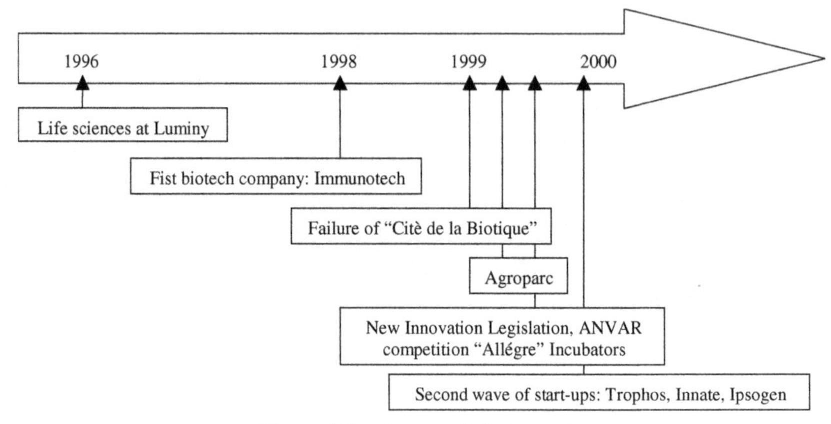

Fig. 6.1: Timeline of the development of the Marseilles cluster.

an economic development agency of the department "Bouches-du-Rhône" (BDR), created around years 1990 with 18 experts, to help both French and foreign companies to establish themselves successfully in the region.

Currently the cluster is growing very slowly and the number of employees is still modest. Entrepreneurs are more concerned with "creating their own job" rather than with strengthening their companies competitive position and thinking to enlarge the scale of the firms (Fig. 6.1). The same "conservative" approach may explain why no company has tried to go public in the past, besides of course the present situation of the public market, which is totally antagonistic to such a project.

6.2 Major Actors

The analysis takes in account: (i) the DBFs; (ii) the industrial and research environment; and (iii) the financial environment.

6.2.1 Dedicated Biotech Firms

Overview

Table 6.1 presents an overview of the DBFs in the cluster of Marseilles. The cluster is still in its embryonic phase. The industrial base, indeed, does

Table 6.1 General information on the Marseilles cluster.

Number of DBFs present in the cluster	12 (+ 4 in Aubagne/Gemenos' region)
of which local DBF	11
of which foreign DBF	1
Number of public firms	0
Number of profitable firms	5 Biotech Germande, Biotechna, Gelyma, Germe SA, Immunotech
Total turnover of the DBFs of the cluster	€38.2 million (2001) (*)
Total employees of the DBFs of the cluster	400
Average size in terms of turnover	€3.18 million (more representative is €0.56 million = average of the 11 companies (without Immunotech))
Average size in terms of employees	33 (200 for Immunotech and 18 on average for the others)

(*) €32 million for Immunotech and €6.2 million for all the others.

not present large companies active in the biotech sector and includes only few DBFs, with no public firms. Among these companies, as previously mentioned, Immunotech represents a successful exception both in terms of turnover and in number of employees.

DBFs were mainly funded during first years of '90s. In the latest years (1999–2002), only three new DBFs were founded by people who left their former company, Immunotech after it has been bought by Beckman Coulter. Moreover, there are no public companies within the cluster. As far as the main field of application is concerned (Table 6.2), clearly it appears a predominance of companies operating in the healthcare sector. Among these firms, however, only 4 (Dynabio, Innate Pharma, Natural Implant and Trophos) develop therapeutic products or kits, while the others are better concerned with platform technologies or services. This may be a result of the mentioned "conservative" approach of entrepreneurs. Indeed, business models related to the development of therapeutics are characterised by a higher risk than the others and require a longer time horizon.

Table 6.2 Main fields of application of DBFs in the Marseilles cluster.

Main fields of application	Number of firms
Healthcare	9
Agro-food	0
Nutraceuticals	0
Veterinary	0
Environment	2
Others	1 (cosmetics)
Total	12

Turnover and employment profile

As for an embryonic cluster, turnover is very low (Table 6.3). Except Immunotech, the other companies have an average turnover less than €3 million, with 72% of DBFs that do not reach one million per year.

As a consequence of the experimented level of turnover, also the R&D expenses are in most cases worth less than €1 million (Table 6.4). Moreover, the investments in research activities concern the companies' business models. In the case of the Marseilles cluster, as noted, the large majority of companies are involved in the development of technological devices or services: Immunotech, for example, spends in R&D near 10% of its turnover, while worldwide biotech companies involved in the development of therapeutics on the average devote more than 25% of their turnover to R&D.

Table 6.3 Turnover of DBFs in the Marseilles cluster.

Turnover	% of firms
Below 1€ million	72
€1–2 million	9
€2–3 million	9
€3–4 million	—
€4–5 million	—
Beyond 5€ million	9

Table 6.4 R&D expenses of DBFs in the Marseilles cluster.

R&D expenses	Number of companies
Below €1 million	11
€1–2 million	
€2–3 million	
€3–4 million	1 (Immunotech)
€4–5 million	
Beyond 5€ million	

Table 6.5 Number of patents in the last three years in the Marseilles cluster.

	Before 1999	1999	2000	2001
Registered patents	34	7	8	3

The analysis of the patents registered by the DBFs in the cluster of Marseilles shows the results of such low investment profile in R&D. There are few biotech patents registered in the last three years and, moreover, the trend shows a further huge reduction in 2001 (Table 6.5).

Table 6.6 presents the data concerning the employment profile of the companies, showing an average number of employees of 34 per company in 2001. It is, however, strongly skewed by the presence of Immunotech (without it, the average number is 18 employees per company).

Process of foundation

With the relation to the process of foundation, Table 6.7 summarises the results of the analysis in the cluster of Marseilles.

There are no start-ups and the large majority of companies belong to the research environment (universities and research centres). This, on the one hand, may be the outcome of the above mentioned lack of entrepreneurship but, on the other hand, it shows a positive attitude within universities and

Table 6.6 Number of employees (2001) in the Marseilles cluster.

Firm	Number of employees
Alphabio	70
Biocytex	12
Biotech Germande	15
Biotechna	17
Dynabio	<10
Gelyma	3
Germe SA	<10
Immunotech	200
Innate Pharma	20
Ipsogen	15
Natural Implant	10
Trophos	21
Total	403

Table 6.7 Process of foundation of DBFs in the Marseilles cluster.

Process of foundation	% of the total
Start-up	0
Industrial spin-off	33
Academic spin-off	17
Scientific spin-off	42
Joint Venture	8

- Start-up: it is referred here as a firm that has not any formal relations with previous industrial or scientific entity;
- Industrial spin-off: it is referred here as a firm that has some formal relations with an industrial actor (parent company);
- Academic spin-off: it is referred here as a firm that has some formal relations with a university;
- Scientific spin-off: it is referred here as a firm that has some formal relations with a previous research centre;
- Joint Venture: it is referred here as a firm formed by the formal collaboration between two other actors.

research centres in supporting the exploitation of research result through the creation of new ventures. Immunotech itself, who gave the way to the cluster birth, is a scientific spin-off from the Luminy Immunology Centre. There are three cases of industrial spin-offs (Trophos, Innate Pharma and Ipsogen), the consequence of the change of ownership in Immunotech (some top managers decided to keep their autonomy); in the case of Biotechna, instead, the creation of the new company was the result of the corporate strategy of SEM (a medium and well established water supply company). As a totally owned subsidiary, indeed, Biotechna is involved in the selective waste collection, in sorting mud and domestic green and industrial waste, thus contributing to increase the innovativeness of its parent company.

Case studies

Immunotech

Immunotech acted as a starter in the minds of researchers of the Luminy University community, and even of the whole French university community. It showed in a pioneering way that researchers could become entrepreneurs. And this was a sort of psychological and socio-political shock, because there was, at that time, a huge distance between them, with the universities and their professors and researchers on the one side, and the industry and commerce on the other side. The former considered the latter as lower in status and — on an ethical scale — not very valuable. Since the change of government in 1981, however, there has been a slow but deep change in this point of view, industry and business gaining progressively some respectability in the academic and scientific environments. This was strongly related to the excellent scientific reputation of the first pioneers, including the founders of Immunotech.

The biotech company was created to develop and commercialise applications of monoclonal antibodies. It started with 7 people. In 1985 they were around 40, 170 in 2002, and almost 200 currently, with 50 researchers. The company gained its first positive earning result as early as 1987. Between 1986 and 1995, Immunotech created commercial subsidiaries in a number of countries, including the US. The fast growth forced the company to enlarge its perspective: in June 1995 Immunotech became subsidiary of Coulter and in October 1997, merged with Beckman. The resulting company was the third in the world for clinical diagnosis. Immunotech is currently focusing on three main areas with three departments: Cytomics (flux cytometry); Immunoanalysis (kits for immunodosage); and Immunomics which specialises in reagents for the measure of the specific response of T lymphocytes. In 2002, the catalogue offered more than 1,600 of such reagents. The expansion is continuing with a turnover of €32 million (17% of which represents the net benefit after taxes).

Trophos

Trophos started its activities in 1999 in the science park of Luminy, thanks to some of the founders of Immunotech. The aim of the company is to apply a new approach to drug discovery for neuro-degenerative diseases and particularly to develop models of purified neurons in culture. Moreover, Trophos has developed a system of high throughput screening (HTS) to select new therapeutic molecules. Regarding this issue, the major problem for Trophos was to scale up their HTS devices to an industrial scale: recently this result was achieved through the development of a robotized platform and a novel cell analyser.

The source of molecules is at the same time very important for the productivity of HTS devices. Trophos uses innovative chemistry originating in the "Laboratoire de Chimie Moléculaire" from the Faculty of Sciences as well as powerful selection tools such as DNA arrays, predictive toxicology on embryos, and transfected cells, even benefitting from the collaboration with the scientists at the LGPD (Institut de Biologie du Développement de Marseilles).

Trophos, after ending its first round of financing, has recently signed a large contract with "Association Française contre les Myopathies" (AFM) to finance a research program for Spinal Muscular Atrophy. Other projects concern Amyotrophic Lateral Sclerosis, Huntington's disease and Alzheimer's disease. Trophos is now developing rapidly. It currently employs 19 people, including 9 PhDs, in three operational units: Biological Models, Chemistry and Screening.

Ipsogen

Ispogen was founded in September 1999 in the Grand Luminy incubator as a spin-off of Immunotech. Since the early '90s, Ipsogen scientists have been using large-scale biochip gene expression measurement technologies in their labs (in academia or in Immunotech). These technologies have been successfully applied to their chosen areas of expertise (oncology, immunology and gene functions' identification in transgenic mice), thus giving the company a unique position in the field of biomedical applications. Large-scale gene expression analysis provides essential information not only on tumours biology, but also in predicting onset and behaviour of the disease in response to planned treatments. The major outcome of such analysis is to allow the definition of predisposition in a given individual to develop a cancer, whether or not the tumour is aggressive, and which drugs will be effective in killing those cancer cells.

Ipsogen uses its technology platform to identify gene sets that characterise tumour biology at the individual level and allow clinicians to apply existing and innovative treatments more efficiently. Ipsogen focuses on leukaemia and breast cancer, and particularly capitalises on the knowledge of its academic partners. The main strategy of Ipsogen is to industrialise its technology platform such as to engage a significant number of projects simultaneously with the objective of patenting its discoveries and marketing innovative diagnostic tools. A test for leukaemia prognosis, to be implemented in few years, is one of the major goals of the company. In the mid-term horizon, Ipsogen aims at focusing in breast cancer and leukaemia and at implementing a global managed care approach to these diseases (from the predisposition and diagnosis to prognosis and targeted treatment). Ipsogen's technology platform is composed of: (i) ultra sensitive biochips; (ii) large libraries of clinically documented tumour samples; (iii) processing methods of biological information; and (iv) patented processes for gene expression analysis in oncology.

Thanks to its variety, this technology platform is able to generate a continuous stream of products in the field of diagnostic oncology and, more generally, to offer specialised services for the biopharmaceutical companies. Ipsogen staff is currently made up of six employees and the headquarters is located in Luminy, near the Marseilles University campus and close to the major institutes such as the Paoli-Calmettes Institute Anti Cancer Center and the Marseilles Luminy Immunology Center (CIML).

Innate pharma

Innate Pharma, founded in 1999 in the campus of Luminy by people belonging to Immunotech, is a product-oriented biopharmaceutical venture focused on the pre-clinical stage of development. Innate Pharma was the first company to focus on the pharmacological manipulation of non-conventional lymphocytes (gamma delta and NK cells), thus identifying its primary clinical targets in oncology. Further clinical development is contemplated in other therapeutic fields such as infectious diseases, allergy and auto-immunity.

On April 2000, as result of a first round of financing, Innate Pharma raised €4.5 million from the major European venture capital firms. This financing supported pre-clinical (and allowed the first steps of clinical) development of the identified drug candidates. In July 2002, a second round of financing provided to the company further €20 million, funding the development projects of novel anti-tumoral therapies, and significantly "exceeding" the expected €15 million. Co-leaders of the financing are Alta Partners, a Californian venture capitalist, and Axa Private Equity, an investment management company in the Axa group which operates in more than 50 countries. Other investors include Pechel Industries (Paris), Gilde Biotech & Nutrition (Utrecht, Netherlands) and Innoveris (Marseilles). The original investors (including first-round financing leader Soffinova Partners, as well as GIMV and Auriga) have also significantly increased their investments for the second round of financing. The success may be seen as a clear "vote of confidence" in the innovative technology, based on stimulation of innate immunity, used by Innate Pharma. Moreover, these resources allowed Innate Pharma to begin its clinical trials program and to pave the way for a second generation product research programs. Particularly, the products "candidate" for clinical trials include: (i) Innacell-gd, a cellular therapy for certain types of renal carcinomas based on the use of a novel chemical entity activating *ex vivo* a population of anti-tumoral cells (Tg9d2 cells); (ii) Phophostim, an immunostimulatory drug intended for systematic administration that also relies on the activation of Tg9d2 cells for two cancer indications (multiple myeloma and renal carcinoma); and (iii) Kiromab, a cytotoxic monoclonal antibody that will enable the body to fight against certain coetaneous lymphomas (orphan indication) for which no satisfactory treatments currently exist.

6.2.2 Industrial and Research Environment

Historically, universities played a major role in the scientific development of the area of Marseilles. Strong medical, pharmaceutical and biology faculties have been developing both academic and research activities, resulting in a large number of students (nearly 4,000), post graduates and researchers (nearly 3,000), creating a favourable ground for further initiatives. Among them, mixed research institutions were created in the field of immunology or oncology that became leaders in their areas, like the already mentioned CIML (Immunology Centre of Marseilles Luminy) and the Paoli-Calmettes Institute (IPC). Unfortunately, political considerations nearly 30 years ago resulted in a "split" of the universities of Aix and Marseilles (two towns 35 km close to each other) into three universities and the scientific departments and university labs have been spread between these three entities, thus making the speciality "map" of universities quite complicated. More recently, closer relations between academic and research structures are implementing. In many cases teaching resources depend on two or more universities and the accreditation by the Ministry of Education and Research supports curricula of scientists in the name of two or more universities as well. This creates a very complex network which may be explained as a mechanism to overcome (more or less successfully) these artificial and anachronistic "separations". Recently, a new trend emerged concerning the development of relations between academic "sectors" who had no relation at all before or even, till few years ago, ignored or disregarded each other, like scientific units and management units. For a couple of years there has been a growing interest in interdisciplinary curricula and the mixing of students of different educational backgrounds in common academic activities. This is the case especially of academic campuses where universities and other schools (i.e. engineering or management schools) are physically located in the same place. The campus of Luminy, based on the southern side of Marseilles, is one of the earliest examples of this practice and actually plays a significant role in the current profile of the academic structure in the biotech sector. An interesting example is the ISTMP (Institut Supérieur de Technologie et Management, Provence) which originated from a network (the "Réseau ISTM National") coordinated by the Chamber of Commerce of Paris. The Institute, hosted in the ESCMP (Management school of the Marseilles Chamber of commerce and Industry) of the Luminy Campus, offers a 3-year program organised around three main subjects: (i) strategy and management; (ii) science and biotechnology; and (iii) international training. The program leads to a double degree: a diploma delivered by the Chamber of Commerce, and a university degree (Diplôme Universitaire) delivered by the Université de la Méditerranée.

The Region of Marseilles (PACA) is the second French pillar for research in biotechnology after Paris, mainly because of the strong focus of its research centres in Health and Life Sciences. Other domains like agro-food, chemicals, and bio-informatics are also concerned, with more than 3,987 researchers, engineers and technicians working in the PACA Region in the biotechnology sector (around 1/3 of the total number of people in public research). Research is mainly public and consists of more than a hundred public research labs, half of which being located in Marseilles.

Concerning the industrial environment, instead, apart from the case of Immunotech, there are no large companies in the cluster involved in the biotech and pharmaceutical sector and, surely, no traditional pharmas preexist at the cluster birth. This lack of industrial base, which only in part is overcome by a strong research base and an effective and efficient system of incubators and science parks (e.g. the Luminy centre), actually represents a peculiarity of the Marseilles cluster. All the companies in the cluster, indeed, if the "academic origin" of Immunotech is concerned, belong to the research environment.

6.2.3 Financial Environment

The support of venture capitals, either local (Samenar and Sofipaca), national (Sofinnova, Auriga, Turenne Capital, . . .), or foreign (Alta Partners, ABN Amro, . . .), strongly influenced the growth of the companies within the cluster. In particular, venture capitals provided the companies the second round financing, thus enabling them to survive after the initial "captive" development in the incubators and science parks within the region. Besides private equity financing, however, there are no direct public interventions. Indeed, only the three Immunotech spin-offs (Innate Pharma, Ipsogen and Trophos) gained the access to public financing, with an average amount of funds of nearly €1 million. The France government, both at central and local level, prefers to sustain research centres (all the public labs and universities in the region receive a large amount of public funds). In this way, it indirectly allows both scientists to carry out their research with state-of-the-art facilities and devices (resulting in a potential high level of innovation) and research institutions to create large technology transfer structures and to support entrepreneurial activities, even managing autonomously dedicated funds.

6.3 Context Factors

The French government strongly favours the development of innovative sector like the biotechnological one.

The Law on Innovation and Research ("Loi Allègre"), launched on July 1999, had the fundamental objectives of favouring the technology transfer process from public research centres to biotech companies and of supporting the creation of new biotech start-ups. Due to the analysed low level of interaction between industrial and academic environment, federal government decided to "force" collaboration agreements. In the region of Marseilles, the law allowed the birth of new ventures like Ipsogen or Urma R&D. According to this law, researchers, academics, engineers, technicians, and managers from the university or public research centres are allowed:

- to offer consultancy services to companies that do not exploit their research results. This kind of agreements are subjected to some restrictions: for example, the university must give its authorisation case by case for defined companies and missions, and a maximum of two simultaneous consulting missions are allowed;

- to offer long-term consultancy services to companies exploiting their research results, without any hierarchical position in the company, but having the possibility to participate up to 15% to its equity, acting in the quality of independent worker, with a limit of 5 years, renewable;
- to participate in the creation of a start-up, exploiting their research results, with the authorisation of the university and having a status of "detachment" from their university duties, for a period of 2 years, renewable twice (6 years maximum);
- to enter in a company, not exploiting their research results, being temporarily detached for a period depending on the particular status of the researcher.

Moreover, the law provides taxes reduction schemes to those companies developing innovative activities (particularly the ones that collaborate with research centres) and simplifies the juridical form for new high-tech ventures, extending rules of PLCs companies also to some kind of LTD companies.

Besides the Law "Allègre", there have been some specific regulations concerning the IPR management in the biotechnological sector. In France, the legal situation is rather complex due to some contradictions between some articles of the law on bioethics, forbidding the patenting both of the elements and products of the human body and of a simple DNA sequence, and the article 5.2 of another "directive" (98/44), also supported by the vote of the Assembly in January 2002, specifying that isolated elements of the human body can be patented. The use of embryo cells for research purposes remains a controversial matter, the former government having suggested to open the possibility to use, for research purposes, embryos (so called "surnuméraires") no more considered by their parents in a birth project, if there was a medical aim. This suggestion is still under discussion, since the beginning of 2003, at the Assembly.

The Loi Allègre strongly contributed to create a favourable legal framework for the development of biotechnology. However, the French government implemented a more "infrastructural" supporting initiative: the creation of Genopoles. Founded in 1998 by the federal government, the Genopoles are complex structure aimed at better supporting the improvement of biotech activities in some French localities (among them Evry, Marseilles,...). The main idea of Genopoles is to exploit the benefits of clusterisation around a central actor in a defined geographic area. The identified objectives for such actors are the following: (i) supporting the research base through the promotion of research activities; (ii) supporting the creation of new biotech firms by providing incubators and other related services for the early development phases of biotech companies (scientific advisor, business plan coaching,...); (iii) managing the logistic elements and supporting the establishment of new academic and research centres; and (iv) coordinating the activities with all the Genopoles centres. In 2000, the whole initiative's budget consisted of €86.5 millions, the great majority of which was given by the federal government. The Marseilles Genopole accounted for €29.5 million and currently includes 25 research laboratories and efficient mechanisms to support the exploitation of research results through the creation of new companies.

6.4 Conclusions

The limited number of actors in the area of Marseilles highlights the fact that the cluster is still in its embryonic phase. The industrial base, indeed, does not present large companies active in the biotech sector and includes only 12 DBFs (with an average number of employees of 18) and no public firms. Among these companies, Immunotech represents a clear successful exception, both in terms of turnover and in number of employees (200). If Immunotech, founded in 1982, is not considered, DBFs were mainly founded during the first years of 1990s. In the latest years (1999–2002), only three new DBFs were founded by people who left their former company, Immunotech, after it has been bought by Beckman Coulter. As far as the main field of application is concerned, clearly there appears to be a predominance of companies operating in the healthcare sector. Among these firms, however, the large majority is focused on platform technologies and only 4 companies actually develop therapeutic products. The cluster of Marseilles is rather peculiar if the process of foundation of its DBFs is concerned. There are no start-ups and the large majority of companies belong to the research environment. Immunotech itself, who gave the way to the cluster birth, is a scientific spin-off from the Luminy Immunology Centre.

The history of the cluster shows that a favourable background for the biotech industry can be traced back to the long tradition in life sciences of universities and research centres. Since 1968, life sciences had been developed particularly in the Science park of Luminy and led to the creation of a relevant scientific base in the field of medicine, pharmacology and biology. However, the actual trigger factor, as previously mentioned, for the birth of the cluster was the birth of Immunotech in 1982. The spin-off mechanism from a research centre was at that time totally unknown in France and the successful results of Immunotech rapidly increased the interest, particularly form the academic side, to such kind of exploitation of research.

Despite the positive aspects of this first "model", there were no immediate followers; there were many "enthusiastic" observers but few of them tried to replicate the pioneers' experience. In order not to lose the "positive thinking" about the biotech sector, dedicated initiatives were settled up, particularly from public actors. Among the others, the major driving forces in the cluster development are:

- the direct intervention of the French government in funding research-supportive infrastructures (Genopole);
- the enhancement of technology transfer mechanisms;
- thc presence in the area of dedicated infrastructures.

The Genopole phenomenon in the case of Marseilles had a strong impact in renewing the interest of existent dedicated institutions (first of all of the science park of Luminy) rather than in creating new centres of development. The objective to fund research-based structures resulted, on the one hand, in a positive contribute to the creation of an excellent scientific base. On the other hand, it represented a stimulus for the financial environment in sustaining the pre-seed and seed financing, effectively allowing the commercial exploitation of research results. As in a "virtuous" circle, the need to efficiently and effectively link together the research and the financial environment forced the implementation of different technology transfer mechanisms (particularly Technology Transfer Offices within universities and research centres).

The development of the cluster, therefore, was mainly due to the presence of public centres for technology transfer (Science Park of Luminy and Genopole of Marseilles). They both act as biodevelopment companies in: (i) "scouting" and supporting researchers willing to found a new company, through the analysis of researches carried out in universities and research centres; (ii) supporting the management in the concept of the idea; and (iii) hosting companies in their embryonic phase and acting as incubators. The central government (in the case of the Genopole) and the local public actors (in the case of the Science Park of Luminy) actually played a pivotal role in the cluster development, effectively encouraging and supporting the original entrepreneurial spirit that led to the birth of Immunotech.

7 The Cluster of Milan

7.1 History of the Cluster

The biotechnology industry in Italy is still in its infancy. Recent reports show that the number of biotech companies in Italy is rather low (54 firms). If a restrictive definition of biotech company (i.e. small-medium firm with a strong research activities, founded after the '70s and using innovative biotechniques, such as molecular biology, genomics, proteomics ...) is considered, this number further decreases to 29. About half of the companies are located in Lombardy and especially around Milano. Therefore, although the development of the industry is very low, there is a point of concentration of the biotech activities and we can talk of an Italian cluster in biotech. The main focus of the cluster, both at industrial and scientific level, is on red biotechnology related to the healthcare sector (pharmaceutical, diagnostic, ...), whereas few actors are dedicated to the agro-food biotechnology.

The cluster of Milano is at an embryonic stage. The biotech activities can be still traced back to the traditional chemical and pharmaceutical base present in Lombardy since the '50s and related to "established firms and public research centres". These firms were mainly the large enterprises, which include the Eni group and Montedison, as well as some mid-size pharmaceutical companies. For Montedison, all biotech research started in its pharmaceutical subsidiary, Farmitalia Carlo Erba, whose main R&D labs were later acquired by Pharmacia. Within Eni, research activities in biotech were developed at EniRicerche and Sclavo Pharmaceuticals. Only few results were achieved, mainly in R&D and in the foundation of a "biotechnology lobby", which tried to better support the industry. As a result, in the '70s and '80s biotechnologies were used in large companies operating the chemical and pharmaceutical businesses. At the same time, however, biotechnology played only a secondary role, employing few internal resources (and with scarce governmental funding) in the research done at universities and public research centres, thus resulting in a poor contribution to the broader development of the sector. In the late '80s few new firms were founded; most were operating low risk business models, such as diagnostic kits and services to large chemical and pharmaceutical companies that actually represented the only "market" for industrial biotechnology applications. All the biotech activities practically halted during first years of '90s, when the Italian chemical sector suffered a crisis, due to the small size and low propensity to R&D, and important scandals occurred in major players during those years. The majority of these firms were then acquired by foreign big actors (e.g. Carlo Erba by Pharmacia), along with many of the biotech projects that were stopped.

On the other hand, scientists at research centres had poor relations with the industrial context. This brought a slow development and a low orientation to application. Furthermore, government supports were very poor (only a few billions euros) and were mainly directed to basic research, thus worsening the problem of the lack of entrepreneurial spirit of the scientific base.

In the late '90s, things slightly changed. At the international level, the pharmaceutical industry faced a crisis that led many large groups to merge, to acquire other companies and to rationalise their activities. These processes of M&A (Merger & Acquistion), as well as the not so favourable general context, made foreign firms move away new R&D centres, thus leading to the closure of many facilities and industrial R&D labs also in Italy. As a consequence, local managers started new entrepreneurial ventures, buying out the labs and facilities to be dismissed (in most cases "at low price"). The New Economy, which drove big amounts of capitals into high-tech sectors, including biotech, strongly facilitated the creation of some spin-offs. Two firms were quoted on the Nuovo Mercato (the Italian Stock Exchange dedicated to high-tech companies).

No actors play a central role within the cluster and/or coordinate internal initiatives or public events.

7.2 Major Actors

The analysis takes in account: (i) the DBFs; (ii) the industrial and research environment; and (iii) the financial environment.

7.2.1 Dedicated Biotech Firms

Overview

Table 7.1 gives an overview on the cluster of Milano. The number of DBFs is quite small (12), and so are the number of public firms. Only one foreign biotech enterprise, NicOx, located their R&D activities in the cluster. All the others are Italian firms. This means that the context is neither renowned nor attractive to foreign realities.

DBFs currently account for 503 employees (more than 40 people per firm on the average) and a total turnover of around €31 million. These statistics are very small compared to other European realities; on the other hand, however, they look like those of German or French clusters seven years ago. If the average number of employees is considered, data are encouraging as it is higher than other major EU realities (e.g. Munich has an average number of nearly 20 employees per firm). The reason is that the industrial structure is rather peculiar. The companies of the cluster were mostly founded as industrial spin-offs by people exiting from large companies. Therefore, the companies had already a certain number of employees from the beginning. On the other hand, there are no academic (or research centre) spin-offs: the typical start-up research firm based with very few employees does not exist in the Milano cluster. Of the three public firms (Biosearch Italia, Novuspharma and NicOx), only one is foreign. In fact, NicOx, quoted at Nouveau Marche in Paris, is a French company, with the headquarters in Sophie-Antipolis (Nice). However, it was founded by Italian managers. They were forced to locate the company in France as the main funders (Apax and Sofinnova) requested it. The two Italian listed companies were

Table 7.1 DBFs in the Milan cluster.

Firm	Year of foundation	Business model
Areta International	1999	Intermediate products such as monoclonal antibodies, polyclonal antibodies
Axxam	2001	Technology platform for the drug discovery
Biosearch Italia (now Vicuron Pharmaceuticals)	1996	Development of therapeutic drugs: products are developed until phase III, and then are generally licensed to mid-size US or Japan firms in order to keep the commercialisation rights in EU. Production of intermediate reagents and out licensing of basic platform technologies
BioXell	2001	Development of therapeutic products until phase III and then out-licensing
Clonit	1987	Development and production of diagnostic tools based on PCR technologies, sold or licensed worldwide
Kerios	2001	Development of therapeutic products in their early research phases, then assigned to other subsidiaries of the group Intermediate products
MolMed	1997	Development of therapeutic products
Newron	1999	Development of therapeutic products to be out-licensed
Nikem Research	1996	Services related to the lead optimization of potential drugs. Out-licensing of platform technologies
NicOx	2001	Development of potential drugs until phase III, then out-licensed
Novuspharma (now Cell Therapeutics)	1997	Early research and development of products to be out-licensed in US and Japan. Direct commercialisation in EU
Primm	1990	Intermediate products (peptides, monoclonal antibodies) and services

founded as MBO (Management-Buy-Out) from large foreign companies, respectively Marion Merrel Dow and Boehringer Mannheim, who were closing or reducing their R&D efforts in Italy. Taking advantage from the high-tech boom Biosearch Italia and Novuspharma went public in 2000. In 2003 Biosearch Italia merged with the US biotech company Versicor, creating a new business entity named Vicuron Pharmaceuticals. Novuspharma announced a merger with the US company Cell Therapeutics.

The majority of firms is concentrated on the development of therapeutic products and on services. Few firms expect to directly commercialise their products and, as expected, the large majority license out their findings to other firms. If the profitability is considered, only four companies (Areta International, Axxam, Clonit, and Primm) have already achieved the break-even. All profitable firms are platform or service oriented companies.

Figure 7.1 shows the evolution of the number of firms, the number of public firms and the number of profitable firms.

As mentioned before, the cluster developed in the second half of the '90s, in which new realities emerged from the restructuring of the pharma industry and was easier, as a result of the New Economy boom, to access to large capitals. Moreover, two places emerge as point of concentration

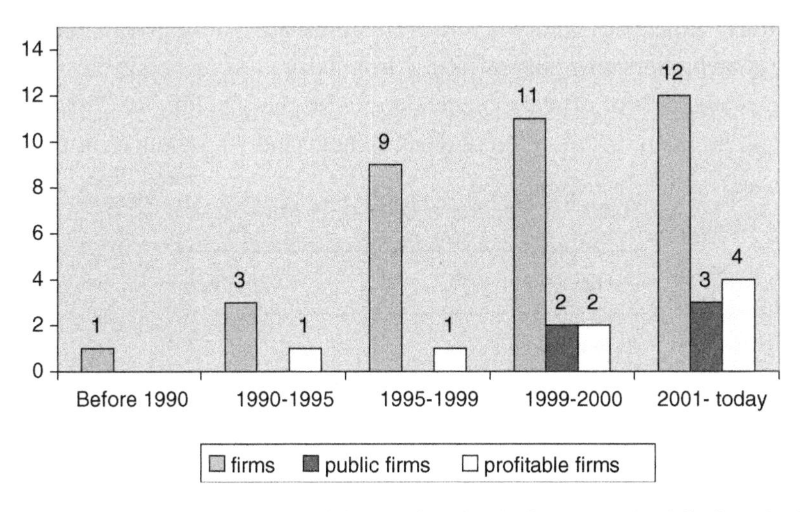

Fig. 7.1: Evolution of the number of firms, of profitable firms and of public firms in the Milan cluster.

of biotech firms: the Science Park Raf (Axxam, BioXell, MolMed, Primm; there was also GenEra, which was acquired by MolMed in 2001) and the former Lepetit area (Biosearch, Newron and Areta).

Turnover and employment profile

The turnovers are rather low: more than 60% of the companies have less than €3 million of revenues (Table 7.2).

This is mainly due to two factors:

- firms focused on the development of therapeutic products are long-term oriented and currently have low revenues;
- firms focused on other businesses (such as services, production of intermediate products, diagnostic kits, etc.) have a very local business and again low revenues.

Figure 7.2 shows the trends in turnovers and R&D expenses in the years 1999–2002.

The figure above shows that the total turnover grew in the last four years as well as that a fundamental element of DBFs is their strong commitment in research activities (about 70% of biotech firms invest more than €1 million per year in R&D).

R&D expenses overcome turnovers of about €10 million in 2001 and have gone higher and higher in 2002, particularly in correspondence with the progressive flow of drug products along with the pipeline. In 2002, there were 8 products in pre-clinical development and 17 products in clinical

Table 7.2 Turnover of DBFs in the Milan cluster.

Turnover	% of firms
Below €1 million	37.5
€1–2 million	12.5
€2–3 million	12.5
€3–4 million	0
€4–5 million	12.5
Beyond €5 million	25

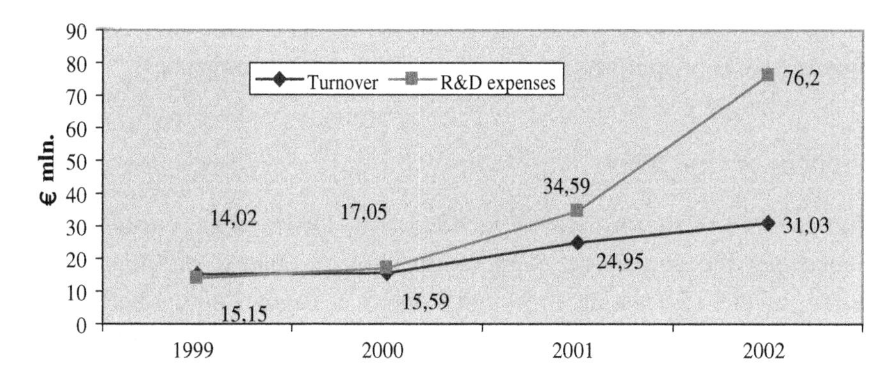

Fig. 7.2: Evolution of turnovers and of R&D expenses in the Milan cluster.

development (8 in Phase I, 7 in Phase II and only 2 in Phase III), but till now no products are on the market, thus highlighting the relative "youth" of the biotech cluster in Milan.

The employment profile of the industry, which is rather peculiar to the case of Milan, is typical of more mature clusters. Indeed, the average number of employees per company (41) is consistently high (Table 7.3).

Table 7.3 Number of employees (2002) in the Milan cluster.

Firm	Number of employees
Areta International	10
Axxam	45
Biosearch Italia	110
BioXell	31
Clonit	3
Kerios	26
MolMed	51
Newron	27
Nikem	40
NicOx	62
Novuspharma	78
Primm	30
Total	503

As mentioned above, however, this is due to the fact that the companies are mostly MBOs or spin-offs from large pharmaceutical companies.

Process of foundation

The analysis of the process of foundation of DBFs in the cluster helps understand the current situation of the biotech context of Milano. As noted, of the 12 existing firms only one is a foreign one, whereas all the others are local enterprises founded in Milano. This is undoubtedly a negative aspect, as it shows the low attractiveness of the cluster for foreign companies. The main problems or obstacles to the location of activities are: (i) the low development of biotech activities in the region; (ii) the rigidity of the labour market; (iii) the absence of incentives which attract new companies; and (iv) the lack of marketing actions and support by governmental agencies at both regional and national level. The process of foundation of all companies are briefly resumed in the Table 7.4.

As far as the process of formation of the DBFs is concerned, the most frequent is that of the industrial spin-off (a company that is founded with formal relations with a parent company) or that of start-ups founded by ex-managers of pharma companies. The total lack of any academic or scientific spin-off should be emphasised, as that represents the main mechanism of growth in any developed cluster (Table 7.5).

The majority of the industrial spin-offs came from the process of restructuring and consolidation that affected the pharma industry during the '90s. Many companies dismissed research centres. This often pushed the managers of such centres to start new ventures with the support of the parent companies (these are the cases of Novuspharma, Biosearch, BioXell, Nikem, Newron, Areta International and Axxam). The support was given under different forms: facilities, assignment of human resources, and research contracts (e.g. the initial contract with GSK accounted for nearly 80% of Nikem Research budget).

The lack of academic or scientific spin-offs in the cluster can be explained with the absence of explicit incentives to the creation of such companies. Only recently, local universities defined the rules about the

Table 7.4 Brief history of DBFs in the Milan cluster.

Firm	Process of foundation
Areta International	The firm has been founded in 1999 by two researchers, previously employed in the Lepetit research centre of Milano, after a three years' feasibility study. In its foundation, the firm has been largely supported by Biosearch Italia SpA.
Axxam	Formerly a Bayer research centre. The firm has been created as spin-off in 2001, after the decision of the parent company to improve its flexibility. The parent company has largely supported the process of spin-off through a five-year contract.
Biosearch Italia (now Vicuron Pharmaceuticals)	Formerly a research centre of Lepetit. After the merger between Marion Merrel Dow (acquirer of Lepetit) and Hoechst group the facility was declared a "non-strategic R&D centre". The management decided to create a spin-off, leveraging on its great scientific base and with the support of a venture capitalist. The parent company gave facilities, funds and a two-year contract. The company, after the merger in 2003 with Versicor, became Vicuron Pharmaceuticals.
BioXell	The firm was founded after the Roche decision to close its research centre in the Milano San Raffaele Science Park with a support from three European venture capitalists. Roche gave facilities, a one-year contract and the right of exploitation of some findings.
Clonit	The firm was born in 1987 as a start-up from two researcher of mid-size pharma, which developed technical know-how in diagnostic tools.
Kerios	The firm was previously known as Norpharma SpA and belong to various European mid size companies. In 2001 bio-pharmaceuticals division has been acquired by Kerios, an international firm operating different businesses.

Table 7.4 (*Continued*)

Firm	Process of foundation
MolMed	From a scientific project done at DIBIT of the San Raffaele Science Park, a firm was born in 1997 as a joint venture between the Science Park San Raffaele SpA and Boheringer Mannheim (then acquired by Roche).
Newron	The firm is half way between an industrial spin-off (people came out from the restructuring of the Pharmacia-Upjohn research centre of Nerviano) and a start-up, as the parent company did not give support.
Nikem	The firms was born in 2001 as an industrial spin-off. In fact, after the merger between SmithKline Beechaam and Glaxo Wellcome, it was decided that Milano's research center should be closed. This pushed the local management to the foundation of Nikem. The parent company gave facilities and a two-year contract.
NicOx	The firm was founded in 1996 in Sophia Antipolis as a start-up. The management is Italian. A new research centre was located in Milano.
Novuspharma (now Cell Therapeutics)	The firm was born as an industrial spin-off after the acquisition of Boehringer Mannheim by Hoffmann LaRoche, because of the decision of Roche to close the research centre. The company announced a merger agreement with Cell Therapeutics in 2003.
Primm	The firm was founded as a start-up in 1990 by an Italian researcher who had worked abroad. It is focused on the offer of intermediate products from its foundation.

generation of academic spin-offs and the relation between the spin-off company and the parent university.

There are also factors which act as barriers to entrepreneurship such as: (i) the socio-cultural environment, where a stable work is still largely preferred to a not stable one; (ii) the fact that entrepreneurs with firm failures in their background are badly considered by the business community; (iii) the

Table 7.5 Process of foundation of DBFs in the Milan cluster.

Process of foundation	Number of companies	Companies
Start-up	4	Clonit, Kerios, NicOx, Primm
Industrial spin-off	7	Areta Int., Axxam, Biosearch, BioXell, Newron, Nikem, Novuspharma
Academic spin-off	0	
Scientific spin-off	0	
Joint Venture	1	MolMed

- Start-up: a new company which has not any formal relation with existing industrial or scientific entities;
- Industrial spin-off: a firm that has some formal relations with an industrial actor (parent company);
- Academic spin-off: a firm that has some formal relations with a university;
- Scientific spin-off: a firm that has some formal relations with a previous research centre; and
- Joint Venture: a firm formed by the formal collaboration between two other actors.

lack of dedicated infrastructure of main universities in order to better exploit scientific achievement (technology transfer offices, patent offices, incubators, etc.); and (iv) the lack of financial support. Some of these factors will be better analysed in the next sections.

Case studies

In this section, the case studies of the two largest companies of the cluster, Biosearch and Novuspharma, are described.

Biosearch Italia

Biosearch (now Vicuron Pharmaceuticals) was born in 1996 as a management-buy-out of the main research centre of Marion Merril Dow (MMD), formerly Lepetit. After the merger between MMD and Hoechst, which gave the birth to Hoechst Marion Roussel (HMR), the R&D centre was declared "*non-strategic*". The management decided to create a spin-off. Hoechst Marion Roussell, supported the project, giving the new DBF: (i) the facilities of the centre; and (ii) the majority of the intellectual property rights related to discoveries done in the centre.

In addition to this, HMR contributed to the first development of the firm through a two-year research and production contract (of nearly €14 million).

The assets given at the beginning and the initial contract sustained the firm for two years after its founding. Governmental funds of MIUR (Italian Minister for Research and University) for nearly €10 million in the period 1997–2000 and other funds (€14 million) given by 3I Europe, a venture capitalist, strongly contributed to the fast evolution of the firm, pushing the company to the IPO. This was done in July, 2000, on the Nuovo Mercato in Milano, with a total fundraising of €126 million. In March 2003, the firm merged with US Versicor Inc., creating a transatlantic company (Vicuron Pharmaceuticals) focused on the discovery, development, manufacture and commercialisation of novel antibiotic and antifungal agents for tough-to-treat infections.

Biosearch Italia is focused on the research and development of therapeutics drugs: antibiotics, especially antibacterial and anti-fungi ones. The pipeline currently includes: one product in Phase III (Ramoplanina) and very close to the market launch, one product in Phase II (Dalbavancina), one in the Phase III (Bi-Acne) and various products that did not yet reached the clinical trial stage.

The business model and the source of revenues changed along with the life of the firm. During the first years, the main revenues were from the R&D contract with the parent company HMR. Now, the main revenues come from the licensing out of compounds developed at most until phase II of clinical trials. Therefore the revenues are represented by royalties and milestone payments. Another form of exploitation is that of finding an international partner to co-develop and co-market new products. The criteria followed to select the partner are: (i) small or medium size; (ii) geographic focus on US and Japan context (whereas there is an explicit policy to keep rights of exploitation in the EU context); and (iii) small production capacity (there is an explicit policy to keep control over production).

The main collaborations were those with Genomics Therapeutics and Versicor Inc. (which merged with Biosearch in 2002).

Other sources of revenues include the selling of basic compounds and the fees from the use of the technological platform and the libraries developed by the company (customers include Schering Plough, Bayer and Menarini).

In the upstream phase (new target discovery, target validation), Biosearch Italia cooperates with scientific actors (University of Milano, University of Moscow), and other local DBF (Newron Pharmaceuticals) to enlarge the range of research projects.

The success of the business model is shown by the rapid growth of the firm both in terms of employees (87 in 1999, 110 in 2002) and in revenues (€8.86 million of total revenue in 2001).

Novuspharma

Novuspharma (now Cell Therapeutics) is the second most important DBF in the cluster in terms of employees (85). Novuspharma was born in 1998, as spin-off from Boehringer Mannheim–Roche. The foundation of the firm followed the acquisition of Boehringer Mannheim by Hoffmann La Roche. Hoffman La Roche decided to sell the Italian R&D centre of Boehringer, focused on oncology. The management, with the financial support of three venture capitalists (3i Europe, Atlas Venture and Sofinnova), generated a spin-off company. The involvement of these actors in the equity forced the firm to go public quite rapidly: the IPO was done in November 2000 at the Nuovo Mercato and the total raised was €164 million. The focus of the firm is the oncology field. The pipeline shows two products in the Phase III, one in the Phase II and various products in the pre-clinical phase.

The business model of the firm is the following. Novuspharma in-licences early clinical or pre-clinical projects and brings these projects to Phase III. Companies nowadays burn out about €30 million per year in order to develop their products, at least those that were planned to be marketed in 2004, leading the firms directly to break even.

Within this business model, many collaborations are done with academia and research centres (new target identification, target validation ...) and with platform biotech (lead selection, lead optimization ...) for the up-stream activities and with biotech companies in the down-stream activities. Examples of collaboration partners are:

- the Istituto Mario Negri, the University of Milano and the University of Milano Bicocca, for the research of new therapeutic targets;
- the US National Cancer Institute, to test some new targets developed by the firm;

- Cephalon, in which Novupharma will further test and develop 50 new inhibitors;
- Micromet AG, to develop a molecule with many potential effects on solid tumors;
- Signal Gene, a Canadian firm that gives potential drug candidates Novuspharma will test in first clinical phases.

As a company policy, Novuspharma keeps the right to commercialise products in the EU market, as Biosearch does. The firm in last years has been characterized by a strong growth with revenues that reached €1.5 million in 2001 and R&D expenses around €31.5 million.

On June 2003, US Cell Therapeutics (CTI) and Novuspharma announced a merger agreement.

7.2.2 Industrial and Research Environment

A certain number of both foreign and national large pharmaceutical companies are present in the cluster. However, most large foreign companies have only commercial activities (only 2 out of 23 have R&D labs in the cluster). The commercial presence of foreign big pharmas is due to the importance of the Italian market (the sixth in terms of revenues in the world). Nearly half of local companies carry out R&D activities related with biotech technologies (molecular biology, HTS, ...), even if they mainly operate in niche markets.

The two foreign large companies with R&D activities in the cluster are Pharmacia (now Pfizer) and Schering Plough. The presence of Pharmacia (now Pfizer) was particularly strong, as the facility in Nerviano (Milano) was a corporate centre of excellence in the area of oncology within the group. This centre, which employs more than 3,200 people (800 researchers), covers a wide range of R&D activities from research to development, with €137 million investments in 2001, a pipeline composed by 10 new potential products, and some other molecules in research phases. Given these premises, the social community expressed a lot of concern when Pfizer recently announced that the centre is to be closed. Schering Plough has 672 employees, mostly involved in productive and commercial activities,

32 researchers and an amount of €3 million in R&D investments in 2001. However, the Schering Plough centre, located at the San Raffaele Science Park, is an interesting case of integration of activities with other centres of the Schering Plough group. The products which are initially (pre-clinical or Phase I stage) developed in the US R&D centre of SP are further developed (Phase II and III) in Italy and finally sent back to US for final developments.

The biotechnology academic environment is rather young. Three universities (of which one is private) created degree courses in biotechnology only in the last ten years. Public universities include the Università degli Studi di Milano and the Università degli Studi di Milano-Bicocca. The private university is the Università Vita e Salute San Raffaele, which started its research activities in 1996 and a biotech degree course in 2001. Currently, there are nearly 200 academic researchers in biotechnology in the cluster, and more than 2,200 students in biotech courses. The analysis of the trend, moreover, shows a strong growth rate. Beside the universities mentioned above, there are the CNR (National Research Centre) institutes and other private institutes, which make Milano a high quality research centre, especially in the oncology field. In Milano, nine main research centres (Telethon Institute for Gene Therapy, European Institute of Oncology, FIRC Institute of Molecular Oncology, Istituto Mario Negri, Institute of Agricolture Biology and Biotechnology, Institute of Biomedical Technologies, Department of BioTechnologies, and Stem Cells Research Institute) operate in the area of biotechnology with almost 800 researchers in 2002.

7.2.3 Financial Environment

The governmental funds (particularly from MIUR — Minister of Education, University and Research) played a pivotal role in sustaining the first companies (Biosearch Italia and Novuspharma) in the initial stages of development. The other funding programmes are focused on specific therapeutic areas (e.g. in oncology) or related to fields of application of the biotechnology other than the pharmaceutical one (e.g. industrial processes).

The regional government of the area of Milano seems to be even more aware of the need to support a strategic field like biotechnology. In the last few years, the Lombardy regional government supported the creation of several centres:

- the San Raffaele Science Park and the birth of a sort of biotech pole in the area of Segrate (Milano) where there are the activities of San Raffaele and the institutes of the CNR;
- the centre of excellence for biotech research at the University of Milano-Bicocca;
- the concentration of DBFs around Biosearch at Gerenzano;
- the birth of another biotech pole in Bresso (Milano) where Biopolo is active and NicOx and Novuspharma are located;
- the financing of the Parco Tecnologico Padano in Lodi (in green biotech). The Region financed the project with nearly €7 million, aimed at the creation of the facilities that will host an incubator and some universities R&D labs.

In addition to these supports, the Region is trying to create financial instruments to support the creation of new firms in high-tech industries (proposing to co-fund entrepreneurial initiatives through its financial branch Finlombarda). Finlombarda is acting to support the creation of new enterprises in the high-tech sectors and dedicated funds have been allocated in the area of ICT and biotech. The total amount of the fund for 2002 was €1.5 million and potential entrepreneurs can get up to 59% of their expenditures (maximum of €50,000).

The commitment of venture capitalists is very limited. The total amount of Italian venture capital invested in 2001 was €2.18 million (with 489 investments and 364 objective firms); only 2% was to biotechnology. There is only one Italian venture capitalist dedicated to biotech: Aliceventures, which is mostly active in UK. On the other hand, there were relevant investments of foreign venture capitalists into Italian firms. The following (Table 7.6) shows major characteristics of financing in the cluster of Milano.

7.3 Context Factors

On a central level, the support is very limited and concerns the ministerial funds dedicated to research activities. No strong or direct incentives, like

Table 7.6 Financing in the Milan cluster.

Firm	**Venture capital** *(year) — € million*	**Governmental funds** *(year) — € million [subject]*	**IPO** *(year) — € million [market]*	**Contracts** *typology — € (million)*
Areta International	—		—	Vicuron Pharmaceuticals (research contract)
Axxam	—		—	Bayer (five-years research contract of €29 million) Altri: Altana, Chiesi, Dompè, Gruenenthal, Newron, NicOx, Recordati, Rottapharm
BioXell	(2002) — 22.7 (2003) — 17.0 [MPM Capital, Index Venture, LSP, NIB Capital , P.E]	(2002) — 0.21 [European Union]	—	Roche (licensing) TaiGen BioTechnology (research contract)
Cell Therapeutics	(1999) — 15.0 [3i Europe, Atlas Venture, Sofinnova]	(1999–2002) — 9.26 [MIUR] (1999–2002) — 5.4 [MIUR] (2001) — 0.25 [Lombardia Regional Government] (2002) — 0.25 [Lombardia Regional Government]	(2000) — 164 [Nuovo Mercato]	Roche (licensing agreements)
Clonit		supported by Finbiotec		

Table 7.6 (*Continued*)

Firm	**Venture capital** *(year) — € million*	**Governmental funds** *(year) — € million [subject]*	**IPO** *(year) — € million [market]*	**Contracts** *typology — € (million)*
Keryos		—		
MolMed	(2001) — 8.8 (2002) — 4.6 [European Development Capital Partnership]	(2000–2002) — 0.75 [European Market]	—	Novartis, Takara Bio
Newron Pharmaceuticals	(1999) — 7.2 (2002) — 25.0 [3i Europe, Apax Partners, Atlas Venture]	(1999–2002) — 6.22 [MIUR]	—	—
NicOx	(1996/97) — 8.3 [Apax Partners, European Medical Ventures, Sofinnova]		(1999) — 33.0 [Nouveau Marchè] (2001) — 59.0 [*follow on*]	Merck, AstraZeneca, Axcan, Biolipox
Nikem Research	—		—	GlaxoSmithKline (two-year research contract)
Primm	—	(1998–2002) — 1.91 [MIUR] (1998–2001) — 1.02 [MIUR]	—	
Vicuron Pharmaceuticals	(1996) — 14.0 [3i Europe]	(1997–2000) — 10.65 [MIUR] (1998–2001) — 0.35 [MIUR] (1998–2001) — 2.09 [MIUR] (1999–2002) — 1.99 [MIUR]	(2000) — 126 [Nuovo Mercato]	Hoechst Marion Roussel (two-year research contract of €14 million)

the ones settled up by the French or German governement in Europe, were implemented.

The major problem concerning the context factor is the lack of a well define legal framework. Among the others, critical issues concern: (i) IPRs, and (ii) biosecurity.

The intellectual property rights policies were rather neglected in the past, given the very low propensity of academic researchers to commercially exploit their findings. The law was changed only recently (2001). The current law regarding IPRs states that the researcher working at universities and research centres is the owner of new scientific achievements. This brings to a reduction in patenting activity: seldom is a researcher allowed to sustain the costs and to follow the process of filing personally. Furthermore, Italian laws did not already accept the EU directives on biotech patenting. Globally, the number of Italian biotech patents in the last thirty years is very low (about 1,500 patents). In the opinion of many important researchers in the cluster, this can lead to a crisis of life science activities in the next years.

Concerning the biosecurity, the Italian legislative context seems to be oriented to receive the EU directives (among the others, 90/219/CEE, 90/679/CEE and 2309/93). The most important law was passed in August 2000 in which the commercialisation of four types of GMO maize was suspended (two from Monsanto, one from Aventis and one from Novartis), although the EU commission did not identify any potential danger. This action caused big actors to close all of their Italian research facilities (such as those of Monsanto and Aventis Crop Science in the cluster of Milano). In such a rapid pace context (Canada, US, China) this negative factor may strongly affect the position of the Italian agro-food sector.

7.4 Conclusions

The cluster of Milano is at an embryonic stage. The number of DBFs is quite small (12) and, even if the biotech activities in the area can be still traced back to the traditional chemical and pharmaceutical base present in Lombardy since the '50s, the cluster actually started in the second half of the '90s. New realities emerged at that time from the restructuring of the pharmaceutical industry.

DBFs currently account for 503 employees (more than 40 people per firm on the average) and a total turnover of around €31 million. If the average number of employees is considered, the data reveals the peculiar origin of the cluster: they are indeed higher than other major EU realities (e.g. Munich has an average number of around 20 employees per firm). The reason, however, is that the large majority of the companies in the cluster were founded as industrial spin-offs from large companies. Therefore, the companies had already a certain number of employees from the beginning. Another peculiar characteristic of the cluster of Milan is the total lack of academic (or research centre) spin-offs, despite a growing scientific base.

The majority of firms focus on the development of therapeutic products (acting primarily as licenser) and on services. The only two listed firms (Biosearch Italia and Novuspharma) that expect to directly commercialise their products have recently merged with foreign companies (Vicuron Pharmaceuticals and Cell Therapeutics respectively).

The favourable background to the birth and development of a biotech cluster was actually represented by the strong industrial base. Particularly, the processes of M&A that interested big pharmas like Pharmacia, Marion Merrel Dow, GlaxoSmithKline, Boheringer Mannheim, as well as the internal rationalisation of Roche and Bayer, forced local managers into starting new entrepreneurial ventures. The role played by the scientific base is rather marginal, given their relative youth in biotech and the huge barriers to the diffusion among researchers of entrepreneurial culture.

Starting from these premises, the major driving forces in the cluster development are:

- the support to the outsourcing and restructuring processes of large companies;
- the public funding to sustain new DBFs in their early stages.

In particular, the local and central government implemented actions and funding programmes aimed at fostering the creation of new biotech companies as industrial spin-offs. On the one hand, public actors facilitated the transfer to the new DBFs of facilities and intellectual properties from their parent company. On the other hand, they provided a funding support to local managers in the management-buy-out process as well as in the early stages of development of the new company. The intervention of the public actors actually allowed the birth of the cluster. However, unlike the other important European cases (e.g. in Germany and in France), the public intervention in the area of Milano seems to be the "addition" of single, one-shot, actions aimed at effectively responding to specific problems rather than the result of a strategic intent in strengthening the biotech industry.

8 Other Cases of Biotech Clusters

In the previous chapters, five cases of biotech clusters at different stages of development have been analysed in depth. This chapter aims to provide a larger view of the clusterisation phenomenon in the biotech sector, analysing other interesting cases at worldwide level.

The objective of this chapter is to identify on a larger empirical base the common characteristics among successful biotech regions. Each cluster has its history and relies on a specific social, industrial, scientific and financial background.

The US represent undoubtedly at worldwide level the place where the biotech sector is most developed and where the clusterisation phenomenon is most evident. Here, the major US biotech clusters (San Diego and Bay Area) are analysed. In Europe, besides the cases examined in the previous chapters, interesting biotech clusters are the ones of Evry in France and Munich in Germany. They represent the major results of the public intervention respectively of the French government (with the Genopoles programme) and of the German government (with the Bioregio Contest). A closer comparison with the cases of Marseilles and Heidelberg (Chapters 6 and 4 respectively) may help the reader to fully understand the strategies they rely upon. An example of a cluster started on the contrary without a strong commitment of public actors, and in which there was not a common strategic scheme, is the one of Oxford in UK. Like the case of Cambridge (Chapter 3), here the spontaneous and commercial nature of the cluster prevails.

Besides major nations, moreover, in Europe there are more and more countries supporting the biotech sector as a way to sustain their competitive position in high-tech, most innovative sectors. Such is the case of Biovalley (a tri-national cluster comprising France, Germany and Switzerland) and of Uppsala in Sweden.

For each cluster, the current situation, the birth and development process, the driving forces and the major actions taken by public and/or private actors are briefly analysed.

8.1 The Cluster of San Diego

The cluster of San Diego (California, US) is one of the most important in the biotechnological sector at worldwide level and the second in the US. The industrial base in the biotech sector includes 216 companies operating in the life sciences (particularly in healthcare). On the average the number of employees is 85 per company.

The large majority of companies in the cluster are product oriented, with a great number of potential drugs throughout the pipeline: in 2001, more than 180 new compounds were in the development phase. It is interesting to notice that the innovative momentum of the cluster is still growing, with research activities leading to more than 200 registered patents per year since 1997.

The industrial base is mostly constituted by spin-off companies, both industrial and academic. Academic spin-offs represented "the first step" in the cluster birth and development, while the creation of new ventures from existing industrial actors started later, actually significantly increasing the growth rate of the cluster.

The biotechnological sector in San Diego owes its considerable strength mainly to two reasons: (i) the excellence of the scientific base both of universities (UCSD) and of private research institutes (Scripps Research Institute); and (ii) a sharp entrepreneurial spirit of the private and public sector.

In the development of the cluster, the University of California San Diego (UCSD) played a pivotal role, acting on both sides of research and entrepreneurship. Among the major initiatives the following can be highlighted:

- the *UCSD Connect*, started in 1985 and mainly promoted by the University of San Diego, together with some entrepreneurs and the local government, played an important role in the creation of an organisation dedicated to create a strong connections among actors within the cluster. This has been primary implemented through dissemination actions (conferences, forums, roundtables, ...), and also through the offer of many services to researchers, scientists, new bio-entrepreneurs and students, who wanted to exploit their biotech ideas;
- the *MIP Awards*, a programme that funds new business ideas, and exploits the results of research conducted in the university labs;
- the *Connect Entrepreneur Development*, a dedicated programme that supports the start-up phase in high-tech companies (by offering a service of "informal" evaluation of the company's potential by venture capitalists and large corporations).

Besides a strong research environment, the area of San Diego has a long tradition in the healthcare sector: a great number of large pharmaceutical companies established their research centres in the area. Among them, of particular relevance are the research centres of Novartis and Dow Chemicals. The excellence of the scientific base in the fields of molecular biology and monoclonal antibodies actually had a great impact on the localisation strategies of these large companies. Corporate labs strengthened the link between Industry and Academia, through research collaborations and integrated development projects in the most innovative biotechnological applications. Finally, this favoured a deeper diffusion of entrepreneurial culture among researchers within the University.

The close proximity of Bay Area (the largest biotech cluster in the world), and of Silicon Valley (the largest high-tech cluster) were also important, as they favoured the availability of capitals already used to deal with high-tech industries, thus making easier for biotech start-ups to get funds, especially from the private sector (venture capitalists, private institutions, banks and business angels). Venture capitalists invested in the cluster in the period 1995–1999 more than US$420 million (10% of the total investment in the biotech sector in the US).

The first biotech company in the cluster, Hybritech, was itself a result of the widespread diffusion of entrepreneurial culture among the scientists at University. Hybritech, leveraging the discovery of the monoclonal antibodies technique, was founded in 1978. The successful example of this company forced the University to create a more formal network for the fostering of entrepreneurship, thus establishing the above mentioned dedicated programmes.

The history of the cluster was strongly affected by two major events: (i) the acquisition in 1986 for US$500 million of Hybritech by Ely Lilly (a large pharmaceutical company based in Indianapolis, US); and (ii) the crisis of the military sector after the end of the "cold war" in the early '90s.

The acquisition of Hybritech by Ely Lilly represented a milestone in the history of the cluster of San Diego. When the acquisition was formally completed, many of the scientists that worked for Hybritech, afraid of the loss of their independence (as a part of a large corporation the research centre of San Diego has to follow central strategies), created new companies (industrial spin-offs), leveraging their huge experience in the sector. The strong entrepreneurial culture of those people led to the birth of an industrial base, actually playing as an "innovative engine" for the whole cluster, offering employment possibilities and creating the conditions for the development of further induced activities. The importance of the analysed process can be better understood by looking at Fig. 8.1.

A second key factor was the crisis of the military sector. A great number of highly qualified workers were forced to search for a new job. The government directly intervened to sustain the restructuring process, searching for a different industrial sector in the area with high growth possibilities and a high-tech nature. Given the presence in the area of an excellent base in life sciences, actions such as detaxation or special funds and credits were taken to favour the creation of new biotech companies (and then employment opportunities). These direct interventions helped the cluster re-starting its growth process.

In the same years, a central actor (Biocom) was created. Biocom, founded in 1991, is currently a multi-service organisation that carries out

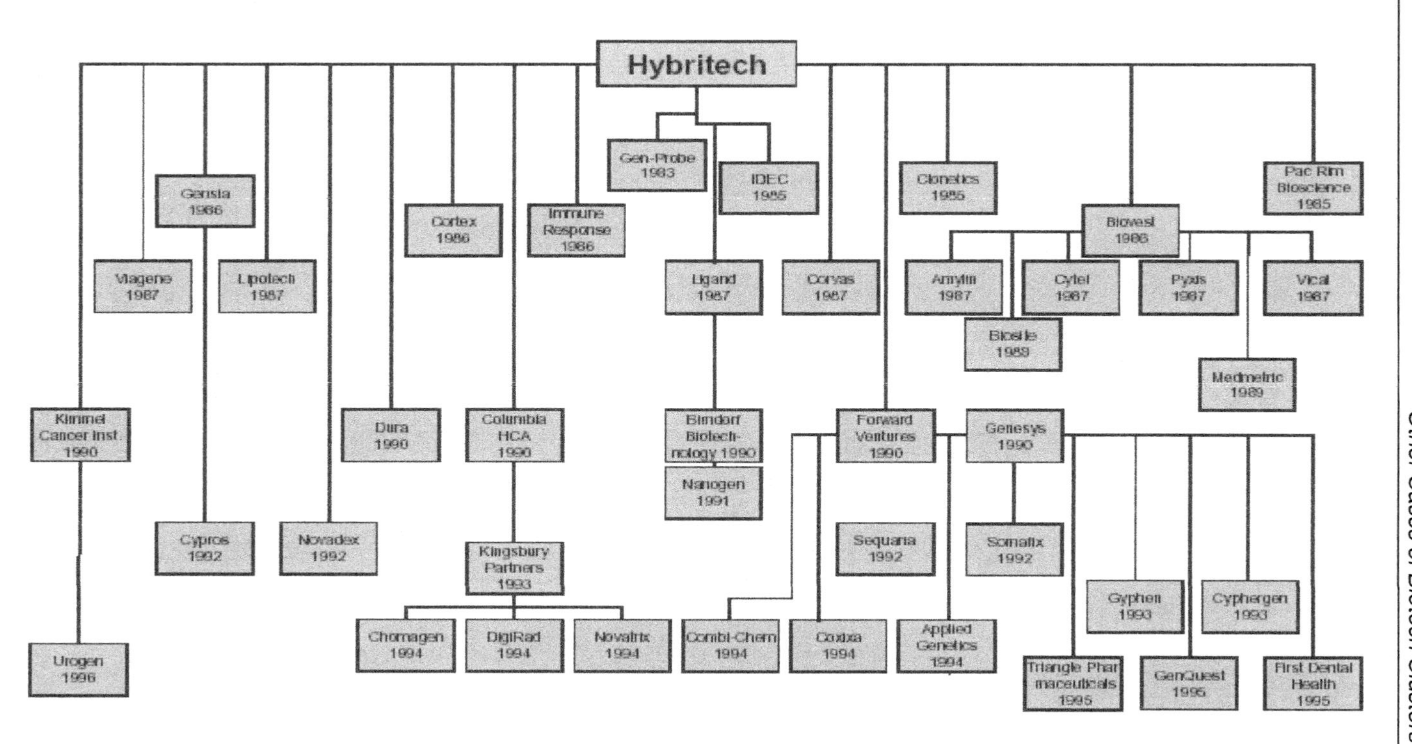

Fig. 8.1: Industrial spin-offs of Hybritech (source: National Research Council Canada).

promotional initiatives and provides financial and industrial services to the companies of the cluster. Its activities include public policy advocacy (through strong lobby actions), participation in purchasing group and promotion of the firms through publications.

The analysis of the cluster birth and development process reveals that the major driving forces are:

- the availability of venture capital;
- the presence of a strong scientific base with a diffused and supported entrepreneurial culture;
- the public intervention in sustaining the restructuring process of the existent industrial base.

The US's favourable and well-defined legal framework, with the Bayh–Dole Act (approved in 1980), as well as the above mentioned proximity to other important high-tech clusters, considerably "leveraged" these basic conditions.

In particular, the Bayh–Dole Act created a uniform patent policy among the many federal agencies that fund research, enabling universities to retain title materials and products they invent under federal funding, and offering the cue for the full exploitation of research results through entrepreneurial ventures.

The cluster of San Diego represents a rather peculiar case. It may be described as a spontaneous cluster where the public intervention (forced by an event external to the biotech sector) actually allowed the cluster to reach its maturity.

8.2 The Bay Area

The cluster of Bay Area, near San Francisco in US, actually represents the "birthplace" of biotechnology. Indeed, in Bay Area the first biotech companies were created towards the late '70s by the academic scientists who put the bases to the modern biotechnology, with the development of genetic engineering techniques. Biotech giants like Genentech, Chiron and Amgen started their operations in the Bay Area. Currently, the cluster encompasses more than 110 product oriented companies (of which 39 are listed at

Nasdaq). However, if the companies focused on diagnostics and medical devices are also considered, the whole industrial base has up to 740 companies. In the cluster, nearly 26,000 people work in life sciences companies, with 2001 revenue from marketed products amounting to US$8 billion. Leading-edge industrial research centres of large pharmaceutical companies (among which, for example, Roche and Bayer) complete the picture of the area.

The large majority of companies are academic spin-offs. The scientific base of Bay Area, indeed, is widely recognised as excellent at worldwide level, thanks to the departments from both of the California University (with three campus in San Francisco, Berkeley and Davis) and Stanford University. Universities actually represent the "innovation engine" of the cluster: they led to the creation of more than 160 new companies in the biotech sector since the origin of the cluster (Fig. 8.2); and royalties gained by the commercial exploitation of their research results amount to more than US$100 million every year. Moreover, universities in the Bay Area are able to financially self-sustain their development, by leveraging the existence of efficient and effective technology transfer mechanisms and taking an equity stake in the generated academic spin-offs.

Initiatives aimed at supporting the development of the biotech sector stimulating the entrepreneurial culture of scientists within universities are constantly implemented.

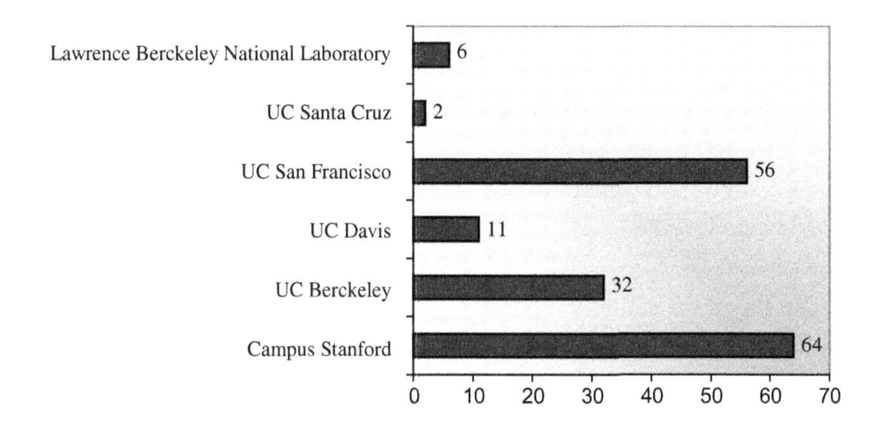

Fig. 8.2: Academic spin-offs in the Bay Area (*source*: Biocom).

Among the others, BioStar may be examined. BioStar is a competition for funding within the California University, based in the innovativeness and commercial exploitability of research projects. The competition, since its start in 1996, has led to the funding of more than 300 projects — totalling US$32 million.

Besides university funding, a huge number of biotech oriented venture capitalists operate in Bay Area. In the period 1995–2001, they funded 261 venture projects for a total amount of US$3 billion. In the same period, 31 companies reached the IPO, thanks to the advisory of local venture capitalists.

The history of the cluster can be resumed as follows:

- thanks to a large scientific base and to the availability of private venture capitals, the cluster started its development in the late '70s. In 1976, Genentech, the first biotech company in the world, was established in San Francisco;
- during the '80s, the cluster experimented a rapid growth in the number of companies, with the birth of 50 new biotech product companies. This growth process led to the creation of nearly 19,000 jobs and the whole cluster in 1987 accounted for a total turnover of US$2 billion. In the same time, scientific research in biotechnology enlarged its focus to other fields of application, particularly in agro-food and environmental biotechnology;
- during the nineties, finally, the Bay Area cluster reached its maturity, exploiting the "enthusiastic wave" concerned with high-tech sectors. This allowed the most successful companies (Genentech, Amgen, ...) to reach a size similar to the one of traditional large pharmaceutical corporations. As a consequence, the cluster established its leadership in the sector at worldwide level.

The Bay Area actually represents the major example of a spontaneous biotech cluster. The public intervention, indeed, only played a small role here. The effective technology transfer mechanisms and the strong scientific base, together with a diffused entrepreneurial culture and innovative funding, led to the creation of the largest biotech cluster in the world.

8.3 The Cluster of Evry

The best way to introduce the cluster of Evry in France is the analysis of its history. Evry represents one of the major example of how the public intervention may force the creation of a biotech cluster. During the '80s and the first '90s the French biotech sector faced great difficulties because of: (i) the theoretical culture of scientists that lacked an application-oriented approach; (ii) the presence of few private subjects able to finance biotech start-ups; and (iii) a burdensome legal and fiscal system.

However, the worldwide development in the mid '90s of the biotech sector convinced the French public institutions to make some changes. They started initiatives aimed at providing fiscal incentives and funding, and created a series of infrastructures, named "Genopoles". The Genopoles are concentrated sets of institutions, universities, labs, and foundations, promoted by public intervention with the involvement also of private organisations.

The Genopole of Evry, founded in 1998, has been the first and the most important of the French realities. It was the result of a joint initiative of public actors (state, regional, and county governments — which showed a strong commitment) and AFM (a private foundation, dedicated to finance institutes and labs working on some particular genetic diseases, that created the first research centre, the Genethon). In particular, the French government chose Evry because of its excellent scientific base in biotechnology and also of its proximity to Paris.

The first step of the establishment of the Genopole was the creation of a science park focused on genomics, able to host also research groups from the near universities of Paris and therefore reach a critical mass in research activities. Moreover, the new science park had strong linkages with the University of Evry and the other major actors (local public actors, venture capitalists and large companies).

The combined effect of these characteristics was to foster the birth of new biotech ventures (start-ups as well as academic spin-offs) commercially exploiting research results. The science park sounded particularly attractive for biotech companies thanks, on the one hand, to its strong scientific base and its active role in promoting technological transfer (through dedicated offices) and, on the other hand, to the services offered (hosting, shared facilities and labs, assistance in financing and patenting).

Finally, the Genopole of Evry, acting as a central actor for the cluster, recently enlarged its range of activities particularly in two directions: (i) promoting the cluster at worldwide level, attracting new biotech companies and key scientists, and (ii) providing funds for companies in their early stage of development. Of particular interest is the *1er Jour* Fund, which provided seed capital for biotech start-ups to the tune of an annual €1.2 million budget.

As a consequence of the strong investments, Evry had a rapid development. In few years, around 30 new biotech firms have been founded and the science base, already recognised as one of the best in the world for its strong commitment to oncology and therapeutic sectors, has been further developed. Currently, 41 biotech companies operate in the cluster (only 10 of which were founded before 1998) with nearly 600 people involved in research activities. The large majority of the firms is constituted by spin-offs from the research centres.

In the cluster, there are 21 world class research centres accounting for more than 1,500 scientists. Among these, the following can be highlighted:

- National Sequencing Centre;
- National Genotipage Centre;
- Infobiogen, devoted to bioinformatics research;
- Genoplant, for the study of genetic techniques to be used in agro-food applications.

Considering the history of the cluster, it is possible to highlight some major driving forces in its development:

- the public funds to research infrastructures (science park);
- the enhancement of technology transfer mechanisms in universities and research centres of the area;
- the availability of seed capital, which allowed new companies to fund their initial research activities.

As previously mentioned, the case of Evry is a major example of direct public intervention. The central government as well as local public actors played a pivotal role in creating the condition to start the process of concentration of biotech companies in the area. Moreover, given the success of the Genopole of Evry, the French Government decided in 1999 to replicate the project in other areas (Lille, Rennes-Nantes, Toulouse, Perpignan-Montpellier, Marseilles and Lyon-Grenoble).

8.4 The Cluster of Munich

The German biotech sector has developed strongly from the mid-'90s as a consequence of the BioRegio Contest (see Chapter 4). This is particularly evident in the Munich case, which represents the most dynamic and important cluster in Germany.

Currently, the Munich cluster encompasses nearly 115 biotech companies that employ more than 3,000 people. Of particular relevance seems to be the fast growth rate of biotech companies. Their number increased in few years from less than 40 (with around 300 employees in total) in the mid-'90s to the current 115 companies. Such growth, though impressive, revealed some weaknesses. Indeed, the flow of capital from the BioRegio Contest, led many new companies to focus on business models more oriented to near-term returns rather than to long-term sustainability. The large majority of firms in the cluster deal with platform technologies and support services: only 12% of cluster's DBFs are specifically dedicated to therapeutic products development.

In the cluster, there are the European branches of most important US biotech companies, several pharmaceutical and chemical corporations (GlaxoSmithKline, Bayer, ...), together with venture capitalists and merchant banks committed to the biotech sector.

Besides a strong industrial base in the biotech sector, the scientific base presents two university hospitals, three Max Plank institutes and the National Centre for Environment and Health.

The sudden birth and development of this cluster is strongly connected with the deep commitment of the German state (both at federal and regional level), which undoubtedly played a key role, re-launching the whole sector through many initiatives. The objective of winning the BioRegio competition and hence the public money forced all the actors of the Munich area to join their efforts strongly and to develop a common strategy. This resulted in a rapid consolidation of all the relationships that had been fairly weak before.

The BioRegio competition boosted the birth of numerous biotech companies (most of which were spin-offs of research institutes, universities, and large pharmaceutical and agro-chemical firms). The Bavaria Government created a special fund (the Bayern Kapital) aimed at supporting new biotech firms and at creating in 1997 a fundamental organisation: Bio$^{\underline{m}}$AG, that became the key actor of the Munich cluster.

Bio$^{\underline{m}}$AG is the main coordinator of all the activities developed in the Munich cluster and this is particularly important because it led all the actors to share a common mission for the development of the cluster. There were further regional and federal initiatives including: the creation of new infrastructures (such as the IZB incubators in 1998 and 2001) and university campuses focusing on biotech fields, but also more funds for research institutions such as the Gene Centre or the Max Planck institutes.

The history of the cluster reveals how the public intervention actually led to the creation of the cluster, particularly acting on the following driving forces:

- the public funding, both at central and regional level, to the creation of new firms. One of the main outcomes of such funding was to stimulate private investors that already operate in the area;
- the integration among different actors, identifying a common development strategy for the whole cluster;
- dedicated infrastructures (IZB incubators), offering services for the new biotech companies.

As in the case of Evry, the deep commitment of governmental institutions in Munich also gave all the actors a greater awareness of their belonging to a common reality inducing them to participate actively to the strategy that the public actors wanted to implement.

8.5 The Cluster of Oxford

The cluster of Oxford (UK), is, after Cambridge, the most important biotech cluster in UK. Currently, 85 biotech companies operate in the area, 60 of

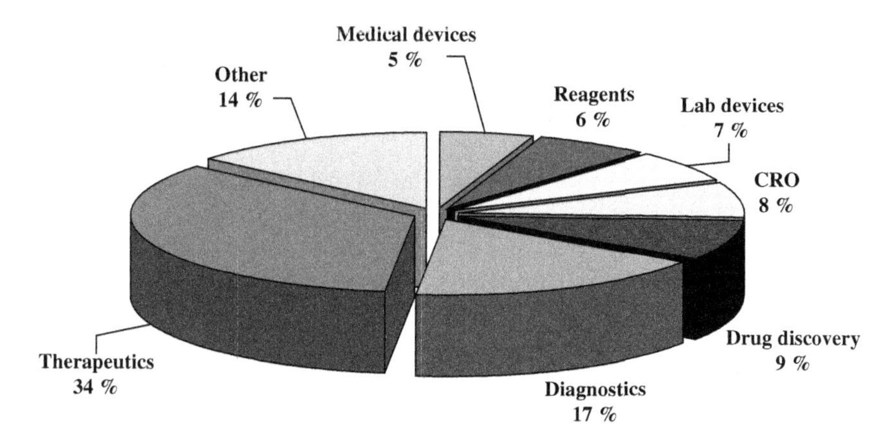

Fig. 8.3: Fields of application of DBFs in the Oxford cluster (source: Oxford University).

which were founded after 1990. The city of Oxford has a long tradition in the sector (the first biotech firms were founded in the early '80s). The majority of these firms is constituted by academic spin-offs, carrying out activities particularly in the development of therapeutics and diagnostics tools (Fig. 8.3).

If service companies are also considered, the number of firms carrying out biotech activities more than doubles. Moreover, there are research labs of major multinational pharmaceutical companies (GSK, Eli Lilly, Merck, Novartis, Abbott and Pfizer) in the area. The attractiveness of the region for such companies relies upon the excellent scientific base, particularly in the field of molecular biology and monoclonal antibodies.

The University of Oxford and the University of Oxford Brookes are acknowledged as centres of excellence at worldwide level for biotechnology. Moreover, large research centres in the area are: the John Radcliffe Hospital; the Central Laboratory of the Research Council (CLRC); the Medical Research Council (MRC); the Radiobiology Institute and the Wellcome Trust Human Genetics Centre. Universities and research centres are strongly involved in the support to scientific spin-offs as an effective technology transfer mechanism. The University of Oxford, for example, created a dedicated company (ISI Innovation Ltd) with the objective of assisting scientists in the start-up phase of their new ventures. The University of

Oxford Brookes, even with a less formal supportive structure, assisted the creation of some of the most successful academic spin-offs (which includes, Powderjet Pharmaceuticals, Oxford Biomedica, and Oxford Glycosciences).

As in the case of Cambridge, the cluster of Oxford has spontaneous nature as well. The key driving forces of the cluster are:

- the diffusion of entrepreneurial culture among scientists, enhancing the commercial attitude of researchers;
- the direct support to academic spin-offs. In most cases, universities help scientists in setting up their new ventures with legal and managerial assistance, and take an equity position in the new company. This mechanism, on the one hand, provides additional funds for entrepreneurs; on the other hand, potentially allows universities to obtain a significant capital gain on their investments;
- the availability of seed and venture capital. Besides direct university funding, a great number of venture capitalists operate in the area, investing in biotech and other small local investors focused on the seed stage financing;
- the diffusion of networking culture, that is the establishment of close relationships within universities and research centres and between these one and existing companies in the geographical area of the cluster.

As in the case of Cambridge (and differently from policy-driven clusters), the need for a central actor, which strengthens linkages among academic and industrial environment, appeared later. In the Oxford cluster, such a central actor was founded in 1999 as a joint initiatives of universities, research centres, DBFs, local and national public actors, and took the name of Oxfordshire Bioscience Network. Its objective is to constantly inform and integrate the biotech community of the activities carried out by the different actors, eventually promoting at international level the major results achieved.

In the Oxford cluster (similarly to the Cambridge cluster), neither strong public intervention nor shared actions among actors were necessary to the birth and development of the cluster. The environmental context presents, already from the beginning, all the critical factors needed for cluster success.

8.6 The Biovalley

The Biovalley is a tri-national biotech cluster located at the cross border of France, Germany and Switzerland, in the Upper Rhine Valley (Alsace, South Baden, North West Switzerland). Centred in the triangle of Friburg, Strasbourg and Basel, the cluster developed recently under the joint efforts of the governments of Germany, France and Switzerland. Currently, 121 biotech companies operate in the cluster, most of which were created in the period 1997–2000, and employ a total of nearly 20,000 people. However, more than 500 actors (life sciences companies, pharmaceutical companies, universities, research centres, technology transfer offices, venture capitalists, public development agencies, ...) actually played (and still play) an active role in the development of the biotech sector in the area, employing as a whole more than 250,000 people.

A favourable background for the cluster birth and development was represented by the strong scientific base in life sciences. BioValley has one of the highest densities of life sciences research in the world, with more than 15,000 scientists, 4 universities (Friburg, Mulhouse, Basel and Strasbourg) and 30 world-class research centres. Of those scientis, 5,000 have a higher academic degree (PhD) and work in life sciences research in 160 academic and/or public institutions, in over 400 research groups. More than 3,000 out of those 5,000 scientists are active in basic research. The quality of the scientific work is best demonstrated by the fact that, in the last 25 years, 5 Nobel Prizes for research in chemistry, immunology and genetics have been awarded to scientists working in BioValley.

In order to effectively leverage the excellence in research and to sustain the commercial exploitation of research results through the creation of new companies, three biotech parks were created: in Allschwil (CH), Friburg (D) and Illkirch–Strasbourg (F).

A great number of industrial research centres of large pharmas (Novartis, Aventis, J&J, Dow Chemical, Eli Lilly, Roche, Dupont, Clariant, Syngenta, Abbott, Pfizer, Sanofi–Synthelab; ...) completes the picture of the area.

Despite the favourable analysed background, no effective signs of clusterisation appeared until the public intervention (forced by an exogenous event) triggered the creation of a biotech cluster in the Upper Rhine Region. The idea to support the biotechnological sector, indeed, was initially conceived by some private German actors at the end of the '80s: George Endress and Hans Briner imagined the creation of a Silicon Valley dedicated to biotechnology in the Upper Rhine Valley. Nevertheless, up to 1996, the related initiatives were rather rare. In 1996, the cluster effectively started when the merger of Ciba and Sandoz into Novartis resulted in more than 3,000 highly qualified unemployed people in the life science sector. As a response, a joint initiative of the regional and local governments, development agencies, universities and private companies led to the foundation of the Biovalley Promotion Team. The Biovalley Promotion Team had the main objective of supporting the creation of a biotech cluster, primarily favouring the spin-off process and the R&D outsourcing from Novartis. The first budget of Biovalley Promotion Team was around €1 million, initially from public funds and then mixed (private and public). In 1997, Biovalley Promotion Team obtained a budget of €2.2 million in the Interreg II Programme of the European Union. In 1998, a new legal structure was created with 3 national associations and a central tri-national biodevelopment agency (Biovalley Company). In 2001, a three-year project was started to privatise the Biovalley Company: the main goal of the new structure is "weaning itself from governmental support and becoming a self-sustaining organisation, while at the same time continuing to promote the growth of life science companies and jobs".

The Biovalley Promotion Team and then Biovalley Company objective was to create 400 new DBFs (mainly industrial and academic spin-offs) and 3,000 new places of employment on a 10-year horizon. Since then the initiatives have gained good results: promotional and networking activities (internet site, extranet, industrial guides, scientific and partnering conferences, education programmes, meeting points with capital providers, ...) have strongly contributed to developing the "consciousness" of being in a cluster; moreover, 41 out of 121 new biotech companies in the Biovalley were directly funded (20 companies in France, 15 in Germany and 6 in Switzerland).

The history of the cluster, given its prerequisites, reveals that major driving forces in its birth and development are:

- the public support to the creation of industrial spin-offs from Novartis;
- the integration among different actors (on a tri-national basis).

Biovalley Company represents the "mean" through which central and local governments implemented their strategies supporting the biotech sector in the area. Acting as a central actor and with a strong political and social commitment, it actually outperformed the task of supporting the restructuring process of Novartis, triggering the start of one of the most active biotech cluster in Europe.

8.7 The Cluster of Uppsala

Another case of a successful biotech cluster is the case of Uppsala, in Sweden. Currently, more than 50 core biotech companies operate in the cluster, most of which are industrial spin-offs with a total of 2,800 employees. Two fields of application dominate the biotechnology in Uppsala equally: (i) the development of methods and instruments for research and development (with platform companies well acknowledged at international level like Pyrosequencing and Amersham Biosciences); and (ii) the development of diagnostics tools (with platform companies as Pharmacia Diagnostic and Q-Med). On the contrary, few companies are involved in the discovery and development of new drugs. Besides these core biotech companies, induced activities area are carried out by nearly 140 life sciences companies, employing 4,700 people.

As in the previous case of Biovalley, the scientific base also represented a favourable background for the biotech sector in Uppsala. There are 900 scientists and 3,500 students. The strong focus of research in medicine and biochemistry at University, however, is one of the main outcomes of Pharmacia's presence in the area since 1950, and of its strong linkages with the academic environment. The presence of one of the biggest and innovative player in the pharmaceutical sector at worldwide level initially contributed to the development of a "positive thinking" about biotechnology. As a consequence, the first signs of entrepreneurial activities in biotechnology can be traced back to the late '80s, when some scientists from the University started new biotech companies aiming at offering technology platforms and services for Pharmacia.

Until the mid-'90s, however, the number of biotech firms was very low and there were no actual signs of the existence of a cluster. In 1996 Pharmacia merged with Upjohn. The planned restructuring of the operations concerned the transfer of the company research away from Uppsala. Actually, the closure of the site did not take place. However, many of the scientists working at Pharmacia, sustained by local public actors, created a network of independent companies (spin-offs). This has led to the emergence of a totally new business structure. Instead of one large, global and profitable operator there are today a large number of small companies. The Uppsala County Administrative Board, the Uppsala University, the Swedish University of Agricultural Sciences, the Uppsala Municipality, the Uppsala County Council and the Chamber of Commerce for Uppsala County, all founded the STUNS Foundation that acted as a forum for coordinating initiatives concerning the development of the cluster. In 2003, a biodevelopment company, Uppsala BIO, replaced STUNS Foundation, with a strategic action plan stating that "by 2008 Uppsala–Stockholm should be recognised as one of the world's five most prominent biotech regions, with a growing competitive industry, leading research and education and a good climate for business and people".

9 The Normative Model

In this chapter, the analysis is done on the basis of the clusters examined in chapters 3–8 with the aim of identifying and examining the main driving forces that enable the birth and growth of a cluster. For each driving force, a brief description is provided and a sample of best practices excerpted from the previous chapters will be given.

9.1 Growth Mechanisms of a Cluster

The birth and development of a cluster can be seen as a virtuous cycle, where a central role is played by the continuous generation of new science-based companies (Fig. 9.1).

In particular, the creation and growth of a cluster rely upon the creation of core biotech companies (start-ups) based on a new idea about new pharmaceutical compounds or technological devices. In many cases these new companies are the outcome of spin-off processes. Two kinds of spin-offs can be specifically identified:

- academic spin-offs, in which the universities or the public research centres play a major role. The idea, in this case, starts in the academic labs and a new firm is created by university people (maybe jointly with business people) in order to exploit it commercially;

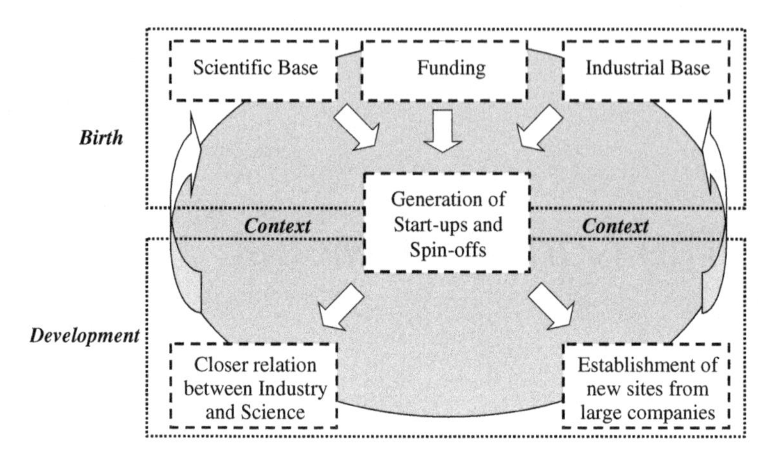

Fig. 9.1: Growth mechanisms of a cluster.

- industrial spin-offs from large pharmaceutical companies. In some cases, these spin-offs are planned by the corporate, which facilitates the creation of independent firms focused on biotech businesses.

It is therefore a matter of evidence that the stronger the scientific and/or industrial biotech base in a geographical area, the greater the chance for a cluster to start and develop there. The process of generating new companies also requires the availability of funding programmes tailored to the funding of new high-tech ventures.

This is increasingly true in biotech, where new biotech companies require large amounts of money from the beginning. Actually, as we have seen in chapter 2, a wide array of business models exists. A common distinction is made between: *product-oriented biotechs*, whose objective is to discover and develop new drugs; and *platform-oriented biotechs*, which are aimed at developing technologies (genomic, proteomic, ...) that support the research process. The former faces a higher risk and a pay-back time that is on average longer than ten years, whereas platform-oriented business models are more concerned with a shorter term and lower risk investment. However, in both cases, the technological and scientific improvements made by biotech-related disciplines and technologies do not more allow "craft-made solutions": the creation of a new biotech firm currently needs more than US$1 million on the average. Hence, availability of funds is a third key factor in a biotech cluster generation.

Finally, a fourth factor is given by the characteristics of the general context: the presence of a favourable "environment" (normative, social, historical, and infrastructural) can actually facilitate the birth and the development of a cluster.

Once the process is started, a virtuous cycle often begins. The strong presence of new innovative biotech companies increases the area attractiveness, facilitating the establishment of new sites (particularly research sites) from large biotech or pharmaceutical companies. The academic origin of some companies, moreover, facilitates the establishment of strong links and networks between Industry and Science. These two effects, in turn, reinforce the industrial and the scientific base of the area and therefore provide the basis for the generation of new ventures (which becomes a factor of attractiveness for financial actors) and so on.

9.2 Driving Forces and Practices

Four main driving forces can be identified, on the basis of the analysis done (Fig. 9.2):

- *financial driving forces*, which concern the availability of funds for the biotech companies;
- *scientific driving forces*, which concern the exploitation mechanisms of scientific research;
- *industrial driving forces*, which concern the exploitation mechanisms of industrial research;
- *supporting driving forces*, which concern the presence of a favourable general context.

The following sections deal with each driving force. For each driving force, a set of factors that enables the driving force to positively act on the area for the birth and development of a cluster is identified. This set of

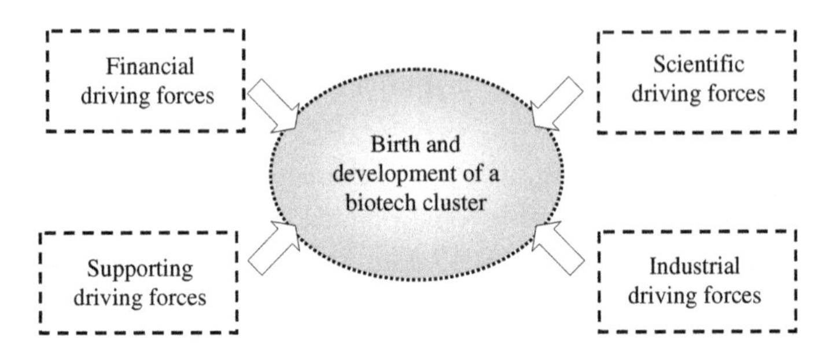

Fig. 9.2: Driving forces.

factors is split into two categories:

- *actionable factors*, which can be directly managed throughout actions made by the actors of the cluster;
- *context factors*, which concern primarily the historical, social and economic background of the geographic area and can not be reasonably modified in the short and middle term.

9.2.1 Financial Driving Forces

Financial driving forces can be identified as in Fig. 9.3.

The scientific and technological advances in biotechnology are expected to drive down costs and reduce time needs by increasing productivity, and to strengthen returns by delivering more effective therapeutics with fewer side effects. Biotechnology is widely recognised as the healthcare paradigm of the future, which is expected to fully change the traditional pharmaceutical chemistry. Despite the great strides made from 1953, however, the process of developing new drugs is still long, risky and expensive. Recent data (Tuft University, 2002) suggest that the average cost of a new drug has risen to US$802 million and that drugs now often require 14 to 15 years to reach the market. Only 250 out of more than 5,000 screened compounds enter the preclinical testing, and only one out of the five drugs that enter clinical trials will be finally approved.

Among the causes, the following can be highlighted:

(i) biotechnology allows to treat more complex pathologies (as, for instance, chronic and degenerative diseases) that, being less known

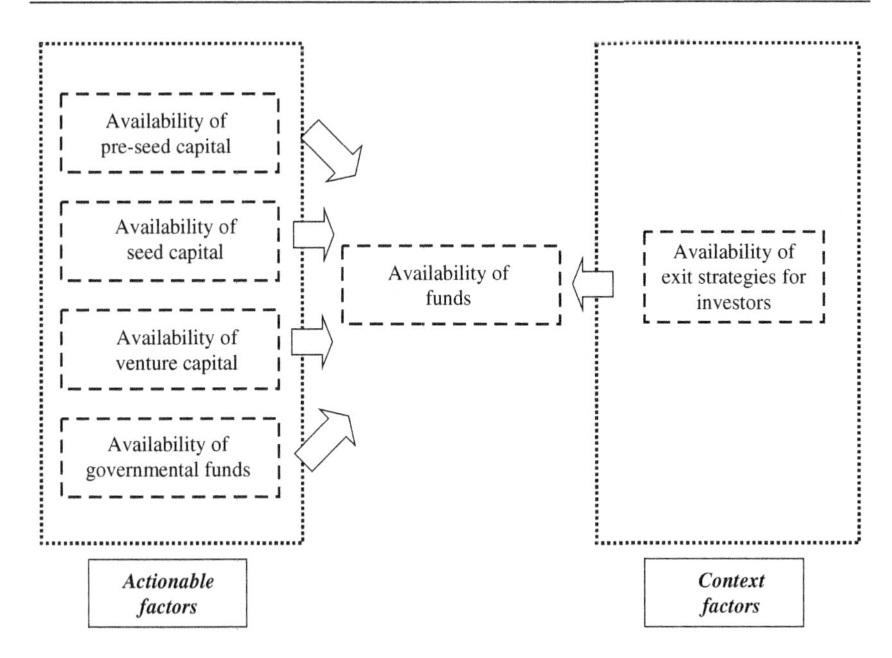

Fig. 9.3: Financial driving forces.

than traditional diseases, cause higher failure rates and longer clinical trials;

(ii) the main technologies are still in the development phase, quite far from maturity, and hence they require increasing investments aimed to achieve state-of-art devices;

(iii) the average approval time of FDA (Food and Drug Administration, the US new drug approval authority) increased constantly since 1998 till now, mainly because of the gap from existing normative and current scientific environment.

The biotech firms already on the capital market have therefore to fight against investors' difficulty in understanding biotechnological business models and, in recent years, suffered from both markets' turbulence and the collapse of the "speculative bead" regarding high-tech stocks.

Moreover, to reach the capital market represents a goal now often foreclosed for the small biotechs. A new biotech firm requires more money, and for more time, to survive longer before it becomes public, in comparison to the '80s when researchers were like pioneers of a new science. Biotechnology is nowadays becoming a basic skill for the sector as a whole, thus increasing the competition around research projects, and, at the same time, technological progresses make lab devices more complex and expensive. Before companies reach the capital market, the main way for a firm to raise capital are private equity investors, especially venture capitalists. VCs are fund managers who invest money into private companies in exchange for an equity position (often relevant). Moreover, even at an earlier stage of development, writing a complete and successful business plan, being able to "catch" the attention of the business community and to demonstrate the commercial exploitation of a scientific idea, requires a wide range of competences and nearly US$15,000–50,000.

Moreover, the main problem of an investor is the way he can disinvest his funds gaining a profit. The most important "exit strategy" is IPO, but recently it has become quite hard to perform, making private investors more careful. There were only three significant equity financing windows (Jain *et al.*, 2002) in the '90s: January 1991–March 1992, August 1995–November 1996, October 1999–November 2000. No other IPO windows have been opened for the biotech sector. Moreover, because of the September 11th attack in 2001, Enrongate and other scandals concerning big corporations, the global economic downturn in 2002 and finally the explosion of Iraq war in 2003, all made financing decisions really difficult.

In comparison to 2000, the financial resources available for the US sector are reduced to less than one third in 2002. Privately held firms have only raised US$1.1 billion, after raising respectively US$3.7 and 3.9 billion in the two previous years. Recent studies (Di Masi, 2001) suggest that almost 35% of pharmaceutical research failures are due to financial problems.

The brief analysis done here shows that the funding of new biotech ventures is rather complex. It should concern the whole life of the start-up from the foundation to the listing on the Stock Exchange and require a number of financial instruments. More specifically, five factors can be identified corresponding to different stages of the company life cycle:

- availability of pre-seed capital;
- availability of seed capital;

- availability of venture capital;
- availability of governmental funds;
- availability of exit strategies.

Availability of pre-seed capital

Pre-seed capital is the capital (on the average less than US$15,000–50,000) which a biotech start-up could use to carry out a proof of concept work and develop a credible business plan. In order to enhance the availability of pre-seed capital, the following practices can be highlighted:

(i) to develop local pre-seed funding and to fund scientists in writing business plans. The traditional scientific education does not provide as managerial competences as needed in order to start a new venture. Particularly the preparation of a credible business plan represents a major problem for entrepreneurs who want to gain access to public and private capitals. After a first screening of the scientific ideas, pre seed capitals, managed primarily by universities or technology transfer offices linked to universities, are assigned to the most brilliant scientists to help them "purchase" the consulting services needed to "transform" their scientific ideas into business ideas;

(ii) to create dedicated support services, like "consulting" offices, particularly within universities. These offices should be able to evaluate the potential for business exploitation of scientific ideas, and to provide managerial and legal competences to plan the creation of new start-ups. Universities that cover a wide range of disciplines could exploit synergies among different departments, stressing the interaction between science-oriented and management-oriented academic people; in this case, instead of directly funding entrepreneurs, the same goal is to make them able to write a credible business plan that is reached through the "inside" offering of dedicated and complementary competences;

(iii) to provide early funds in competition, e.g. business plan competitions for start-ups. This practice gains the two-fold effect of stimulating the entrepreneurship culture among scientists, and of selecting the ideas to fund after the completion of business plans. However, given the amount of funds available in these competitions (on the average less than US$20,000), they represent a way to "reimburse" the winners rather than start a new venture.

Incubation Platform (Evry, France)

The stated mission of the Incubation Platform, a programme of the Genopole of Evry that provides pre-seed capital, is to "identify potential company directors from public or university research laboratories or from private companies, and help them bring their project to fruition". The programme, after an initial validation by a committee of scientific experts, provides the new entrepreneurs with management training and basic information on industrial property, safety, and regulatory standards. In the "pre-creation period" the incubator covers the cost of forming the team of external service providers, and of a marketing strategy consultant, as well as helping entrepreneurs to search for early stage financing.

GSAS Business Plan Competition (Harvard, US)

The GSAS Harvard Biotechnology Club, a non-profit organisation with more than 4,000 members, that hosts events and provides services to explore the world of business and biotechnology operates inside the Harvard University. The Club's mission is to bridge the gap between Industry and Academia by building relationships with companies operating in the biotechnology and healthcare sphere.

Each year, the Club holds a business plan competition for biotechnology related companies. In 2002, the competition was sponsored by DuPont with a US$5,000 first prize and US$1,000 second prize.

Availability of seed capital

Seed capital is the capital (on the average less than US$1 million) with which a new biotech company can actually start-up. The main sources of seed capital are business angels and individuals who invest in private companies, taking over the risk of long-term businesses in exchange for an equity position. An angel, generally, does not have to "answer" to other partners when making an investment (differently from venture capitalists) and is usually less concerned with an exit strategy, allowing companies to work independently also in the event of crisis. At the same time, however, angels are often not sophisticated enough to evaluate biotechnology ventures, and there are only few investors of such typology "dedicated" to biotechnology (particularly in Europe).

Recently, a number of angel networks have emerged, thus increasing the funding capacities of such investors.

In order to enhance the availability of seed capital, the following practices can be highlighted:

(i) development of local seed funding dedicated to biotechnology start-ups. Seed capital directly managed by universities or local organisations (with the sponsorship of private companies, banks, ...) may represent an alternative to business angels;

(ii) support for investors evaluating biotechnology business model. As noted earlier, a major problem for early investors is evaluating biotechnology ventures, given the existing wide array of biotechnology business models. Drug-oriented companies, indeed, face an higher risk and a longer pay-back time than technology-oriented companies, but equally they have bigger expected revenues. Moreover, the relative novelty of biotechnology related disciplines (genomics, proteomics, ...) and their degree of specialisation make the evaluation of new ventures by investors more difficult. External and qualified support from universities or acknowledge public or private association, may help investors in reducing the informative asymmetry and evaluating the scientific reliability of the business ideas. In particular, the "consulting" offices mentioned about the pre-seed capital phase may provide such kind of support;

(iii) creation of (or support the creation of) a network of potential early investors. Often, a major problem in the development of a business idea is the lack of information about funding opportunities. For example, to hold business angels' fairs on the biotechnology industry, or to develop a database of acknowledged backers focused on the life sciences and biotechnology, may help new entrepreneurs to "match" their financial needs.

1er Jour Fund (Evry, France)

Genopole® 1er Jour Fund is a seed capital fund (started in October 1999) managed by the Genopole of Evry, that directly invests in biotech companies being created. Fifteen private and public investors, including a balanced mix of business, banking and institutional actors, have subscribed to the Genopole® 1er Jour Fund for €1.2 million. Future entrepreneurs defend their project against a committee of experts. The evaluation takes into account the work and know-how of the creators, and previously financed projects, which are used as benchmarks. The time span for the contribution to be given back is usually 8 years. This initial capital leads to public financial assistance, in particular provided by ANVAR (the French innovation agency) through Innovation Aid programme, that acts as the financial leverage for the next development stages.

Some successful cases of financed firms are Monoclonal Antibodies Therapeutics (MAT) and ObeTherapy Biotechnology. The equity position of the Genopole Fund in both companies has been recently replaced (together with additional investments) by the iXCore Group (a vc company).

Tech Coast Angels and UCSD CONNECT (San Diego, US)

San Diego Tech Coast Angels is a network of private investors who invest in and assist early-stage southern California companies. It is part of a larger network named Southern California-based Tech Coast Angels organisation, which also has networks in Los Angeles and Orange County, and has invested over US$40 million in more than 50 firms since 1997. Tech Coast Angels offers seed capital in the range of $250,000–$2 million, an investment range that is generally not of interest for venture capital funds. The goal of this network, which is not a fund and in which each investment decision remains individual, is to allow biotech local start-ups to launch their businesses and to accomplish critical milestones that will make them attractive for larger venture capital financing. UCSD CONNECT, founded in 1985 at the urging of San Diego's business community, is the globally recognized, university-based public benefits organization fostering entrepreneurship in the San Diego region that catalyses, accelerates, and supports the growth of the most promising technology and life sciences businesses. Part of the University of California, San Diego (UCSD), CONNECT has a dual role in accelerating growth: it provides added value and delivers targeted, high-level expertise to San Diego's technology business community by teaming up with the region's most prominent industry-specific organizations and individuals, and by partnering with world-class UCSD resources, such as the School of Medicine, Jacobs School of Engineering, San Diego Super Computer Center, and Scripps and Salk Institutes.

In the case of Tech Coast Angels, the UCSD Connect provides a "preferential way" to this network for its assisted firms, thus enhancing their visibility and facilitating their growth.

Availability of venture capital

Venture capitalists are fund managers who invest into private companies. During the '90s, venture capital funds became larger and their minimum deal size has correspondingly increased to the point where they currently does not consider financing of less than US $3–5 million. Generally, venture capitalists do more than just provide money to companies. Differently from business angels, venture capitalists who fund biotech companies have a strong experience in the industry and an established track records helping start-ups to become mature operating companies. Venture capitalists have extensive networks and can help the new entrepreneurs to recruit not only employees, executives, directors, but also customers and other investors. On the other hand, venture capitalists, given their stronger bargaining position, usually negotiate for a large (even the majority) equity position and retain the managerial control of the company, whereas the founders are free to develop their scientific idea.

In order to enhance the availability of pre-seed capital, the following practices can be highlighted:

- to facilitate the access of companies to international funding networks. As noted for the seed capital, and also of greater importance for the venture capital, in which the networking attitude of venture capitalists is more evident, expanding the funding possibilities for the new companies is often a matter of information. To develop efficient "access points" to international funding networks, enhancing the "visibility" of the cluster may strengthen the financial provisions for the companies within the cluster;

- to attract acknowledged biotech-oriented venture capitalists. Venture capital is often the last step before access to the capital market. The current global recession, particularly for the high-tech sectors, enhances the importance for the companies of having a well acknowledged biotech-oriented venture capitalist among their major shareholders. The confidence of the further investors often relies on the "brand-name" of the previous ones.

Financial Forum UCSD CONNECT (San Diego, US)

UCSD CONNECT's Financial Forum traditionally attracts audiences in excess of 400 attendees, including venture capitalists, pharmaceutical and biotechnology executives, investment bankers, private investors, and service providers. More than half of the attending investment audience is from out of state cities including Seattle, Boston, Chicago, New York and San Francisco.

The program combines CONNECT's 12 years of Biotechnology Corporate Partnering experience with 18 years of Financial Forum expertise, both of which have played an important role in the development of San Diego's start-up biotechnology and pharmaceutical industries since 1985. The program provides a "showcase" for San Diego's most innovative bioscience companies and researchers.

The companies are grouped into tracks of 8-minute presentations. All companies have to satisfy three main criteria: (i) they had to have already received venture capital or significant seed investment; (ii) they must be in need of venture funding; and (iii) they must be related to the life sciences industries, including therapeutics, diagnostics, medical devices, drug discovery instrumentation and/or software, and bioinformatics.

Life science presenting companies from the last seven years of the Life Science Financial Forum have raised an average of over US$ 125.6 million in the twelve months following the conference, and deals following the related CONNECT's Biotechnology Corporate Partnering event have concerned more than US$662.8 million since the program's inception in 1988. More than 254 participants representing over 20 bio-industry sectors have participated over the last 12 years of the program. Past presenting companies included DigiRad, Maxim Pharmacueticals, CombiChem, DepoTech (Skyepharma PLC), Recombinant BioCatalysis (Diversa Inc.), Vista Medical Technologies, Alanex Corporation (Pfizer), Innercool Therapeutics, Tandem Medical, and Egea Genomics.

Availability of governmental funds

Governmental funds represent the direct intervention of the local or national government in funding biotech companies. Governmental funds, other than a direct impact, also facilitate the fundraising of companies from private investors, enhancing the companies' credibility. Public funds can be available for all the development stages (from pre-seed to post-IPO). The main problem of such kind of funds, however, is that in most cases the small start-ups fail in "encoding" the bureaucratic mechanisms. Moreover, the *iter* to gain the assignment of public funds is much longer than for private capitals. Therefore biotech companies, which in their early phase of development need a great amount of cash and do not generate revenues, generally cannot wait for public funds and are forced to search for complementary sources of funds.

In order to enhance the availability of governmental funds, and to make it more "suitable" for biotech companies, the following practices can be highlighted:

- to develop governmental "VC style" fund. The "VC style" refers to a less bureaucratic approach and to a closer interaction that is not limited only to the funding process but includes support activities and even managerial support;
- to provide specialised biotech funding programs. Given the characteristics of biotech companies in term of business models (e.g. time horizon, revenues scheme, ...), public actors may create "tailored" funding mechanisms, in which the weight of scientific parameters (e.g. the novelty of the products) as well as of "welfare" parameters (e.g. the therapeutic area of interest) is higher than the weight of the returns' scheme in the near term.

North Carolina Biotechnology Centre (North Carolina, US)
In 1981 the General Assembly of North Carolina created an organisation to stimulate the development of biotechnology: the North Carolina Biotechnology Centre. Initially founded as a state government body, the Centre was reconstituted in 1984 as a private, non-profit corporation, giving it greater flexibility and the role of catalysing interactions between industry, academia and government for technological development. The Centre's budget for the 2001–2002 fiscal year was US$8.7 million and one of its main goals is to foster North Carolina's industrial development through direct investments in new companies.

BioRegio Contest (Germany)
The BioRegio Contest has been the most powerful element in the development of the biotech context in Germany. Launched by the federal government in 1995, the initiative had the fundamental objective of establishing strong linkages between basic research and commercialisation of biotech activities. To build these linkages federal government made available €25 million to those three regions that demonstrated to be in a strong competitive situation and to have great prospective of growth in biotech. The condition for which financings had to be assigned has been particularly important. Governmental funds were available for DBFs requesting financing only if they were also able to collect at least the same amount of money from private investors (regions often gave an additional financial support, creating dedicated biotech funds that concurred to make the private investments one third of the overall investment into a new company). Due to this mechanism, which lowered the risk of the financing (in addition to this, many local contexts gave an additional guarantee that, in case of failure, they would have paid back part of the private investment), venture capitalists "jumped" into the biotech context.

The competition favoured the creation of 17 biotech regions, 4 of which had access to governmental funds. This led in the first two years to the creation of 79 new DBFs.

Availability of exit strategies for investors

The objective of investors, both private or public, is to remove their funds gaining profits in the mid-term horizon (5–10 years), generally through the selling of their equity position in the funded companies. The main way to achieve this objective is to sell the equity position to an other investor (for example, business angels may sell their shares to venture capitalists) and finally to the capital market. As mentioned, the most important "exit strategy" is IPO. It represents for the company the only way to sustain its autonomous growth in the long-term, and for the early investors the easiest way to gain profit from their shares. Capital markets dedicated to high growth stocks, as for the example Nasdaq in the US, enhanced the possibility for high-tech companies to reach the retail investors and to enlarge the set of their shareholders, offering a financial environment more suitable to the businesses of such kind of companies (e.g. with easiest listing procedures) and creating a "customer base" for the companies' stocks more oriented to high risk and long pay back time than the one available in the traditional markets. The creation of a dedicated capital market requires a long period of time and a strong consensus by the central government and a strong commitment of the financial community: both factors go beyond the range of activities that can be undertaken to support the creation of a biotech cluster. Therefore, the availability of exit strategies for investors represents a context factor, rather than a real actionable driving force.

9.2.2 Scientific Driving Forces

Scientific driving forces can be identified as in Fig. 9.4:

In particular, five factors can be identified as influencing the clusters' birth and growth processes:

- presence of scientific base;
- technology transfer mechanisms;
- networking culture;
- entrepreneurial culture;
- mechanisms to attract key scientific people.

The presence of a scientific base is a key driver in the development of a cluster, given that basic research plays a major role in the biotech sector. The creation of a scientific base, however, is part of the history of the area and generally goes beyond the range of activities that can be undertaken to support the birth and development of a biotech cluster. Therefore, it

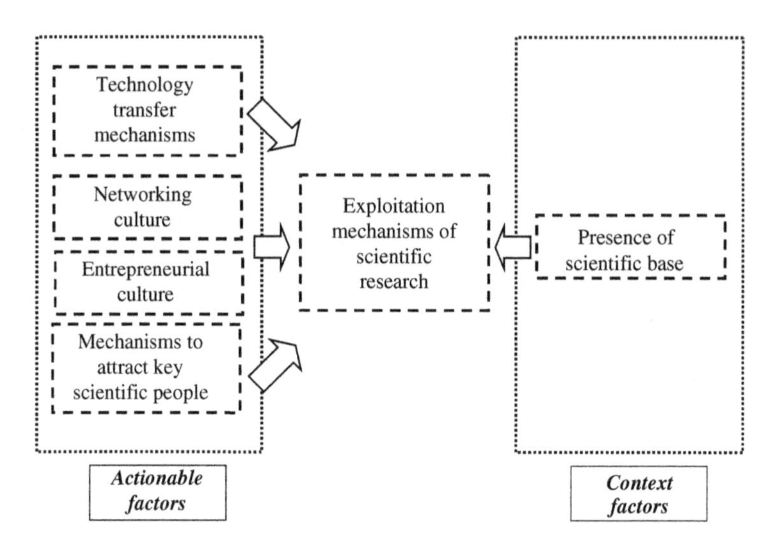

Fig. 9.4: Scientific driving forces.

represents a context factor rather than an actionable driving force. The other factors, instead, can be actually managed in order to leverage the existence of a strong scientific base (that is not sufficient itself) for the development of a cluster. Only efficient and effective exploitation mechanisms of scientific research lead to the creation of a "virtuous" circle.

Technology transfer mechanisms, indeed, on the one hand, allow scientists to commercially exploit the results of their researches primarily through the licensing to existing companies, on the other hand, allow existing companies to "feed" their innovation processes with state-of-the-art technologies and procedures. Through their technology transfer offices universities and research centres can leverage academic knowledge and make it more easily accessible to companies. The commercial exploitation of academic knowledge enhances the reputation and the competitive position of the scientific base. As a consequence, universities can access new sources of funding as well as other forms of pay-offs and are "forced" to set up "internal mechanisms" within their organisations fostering the successful commercialisation of their research results.

The development of technology transfer offices (or similar structures) makes use of (and at the same enhances) the networking culture among scientists and their outward looking at industrial needs, favouring both formal and informal exchanges of knowledge and competences. Moreover, closer relations with the industrial base in the same geographic area strengthen the diffusion of the entrepreneurial culture among scientists. Being open to external industrial stimuli helps academics to understand companies' real needs, the potentials and modalities through which academic research results can be exploited. The diffusion of an "entrepreneurial attitude" within universities and research centres reaches the objective of stimulating scientists themselves in the creation of new companies (academic spin-offs, see introduction). Finally, creating an excellent and innovative scientific environment represents an effective way to attract key scientific people from other areas in the world, favouring the brain immigration towards the geographic area of the cluster.

Presence of scientific base

Common characteristics among the successful biotech clusters include an extensive and successful academic research and education, and a strong industry-academia cooperation. For example, the Cambridge University is widely acknowledged as a centre of excellence, and the exploitation of its leading-edge research in molecular biology was brought to the birth of the cluster. The presence of a huge scientific base is also a main characteristic of the cluster of Heidelberg, where life science research has long been world-class, particularly in the area of molecular biology and immunology.

The presence of leading-edge universities and research centres, indeed, is a key driver in the development of a cluster, making available in the cluster the "lifeblood" of the biotech sector. The scientific base, in order to enhance the geographical concentration of biotech companies, has to reach a "critical mass" in the research activities in all (or at least in the majority of) the biotech-related scientific fields. Regarding this issue, it is a matter of evidence that the task to create a scientific base can be carried out only by a public actor (typically central government) with a long-term time horizon and requires huge investments for the infrastructures and the human resources. Moreover, even if infrastructures can be built in relatively short time, it is quite impossible to establish *ex-abrupto* the excellence in science. Therefore, as noted, the presence of a strong scientific base is the result of the "history" of the area, thus representing a context factor in our analysis.

Technology transfer mechanisms

Leading-edge research focused on the development of new intellectual property is, as noted, a primary driver of the innovative spirit and success of competitive regional clusters. But unless this research can be effectively transferred to the marketplace, the benefit to the regional economy is limited. Technology transfer is the process of finding, creating, and leveraging — whether through licensing or the creation of new products — intellectual property that has potential commercial applications. Such applications are the fruits of research conducted within a variety of research universities and institutions. The autonomy of universities and research centres in an increasingly competitive scenario determines the adoption of dynamic behaviours in the exploitation of research results. In the biotech sector, that is characterised by relevant "scientification" processes, most of research results are easily used by companies as "component/products which are ready for use" and not as "raw materials" which must undergo further long transformation phases. A new discovery in genomics, for example, is directly usable by biopharmaceutical companies. On the one hand, "translating" lab researches into products or technologies may be performed through the direct creation of academic spin-offs. On the other hand, the same results can be achieved through the creation of dedicated structure (technology transfer offices) to "match" the innovation demand and offer.

In order to develop efficient and effective technology transfer mechanisms, the following practices can be highlighted:

(i) to create (or strengthen) dedicated technology transfer offices. These structures, generated within universities, should have a clear managerial independence and maintain a flexible and "thin" organisation (for example with post-doc part-time personnel), thus being able to manage innovative research projects;
(ii) to enlarge the range of activities of existing technology transfer offices, offering a complex set of supporting services, from the general and

project management "consultancy" to the access to both public and private funds. The role of these offices may end in correspondence to the licensing of the project results or to the effective birth of a new venture, eventually maintaining in it a minority equity position;

(iii) to conduct formal analysis to better understand the creation and flow of IP involving faculty, researchers, and students. The objective would be to determine where there are points of leverage to increase both the effectiveness and volume of technology transfer. Local technology transfer offices should be tasked to produce explicit recommendations for the adoption of new IP management-related processes;

(iv) to "match" innovation demand and offer through a stronger involvement of researchers in commercial objectives, "driving" in part the research activity towards the development of useful industrial applications. Moreover, these objectives have to be made available and well-known to (at least) all the actors within the cluster. A strong support is needed for existing networks, both formal and informal, between industry and academia.

UCSD TransMed Program (San Diego, US)

The UCSD Translational Medicine Program (TransMed) has been created to assist scientists in the critical task of moving medical research closer to commercial ready medical technology within the University in order to benefit patients and the public at large. Developed by UCSD Connect, the School of Medicine and the Technology Transfer & Intellectual Property Services Office, TransMed facilitates the access to early funding alternatives for faculty research teams whose work does not fit the model typically funded by federal granting agencies or other traditional funding mechanisms. This program is specifically designed to support promising "translational" research, which is still laboratory based, but is nearing the stage of clinical testing and application. In 2002, 27 high-quality proposals were received by Transmed's board from a wide variety of research teams within UCSD. Six of these proposals were selected for potential funding based upon the "quality" of science being investigated, the capabilities of scientists involved and the extent to which the project was at the "translation boundary", which means it shows potential for validation and possible commercialisation.

Networking culture

Networking culture refers to the ability to create close relationships within universities and research centres and between these ones and existing companies in the geographical area of the cluster. Relations are favoured by localisation, but the process of formal and informal networking may also be strengthened through direct actions.

In order to help the diffusion of the networking culture among researchers, the following practices can be highlighted:

- (i) to support cooperative projects among actors within the cluster. In most cases this allows people from industry to "ripen" products or processes in a close proximity to scientists, whose inputs are useful for further development. Cooperative projects offer the possibility of linking technology, capital and know how to accelerate the technology commercialisation, eventually nurturing new knowledge-based ventures;
- (ii) to promote public and informal meetings. To organise events or conferences where relevant biotech companies and academics are encouraged to interact effectively each other, but even to create "meeting points" as, for instance, restaurants close to academic and industrial sites, may actually keep the flow of technology and business fast and at the commercial edge;
- (iii) to create an environment favourable to networking. Key task in enhancing the diffusion of the networking culture is the establishment of a "positive thinking" among actors, which have to be self-conscious of their belonging to an innovative and competitive cluster. Actions may be taken in order to enhance credibility and trust among actors.

BayBio (San Francisco, US)
BayBio's major objectives are the following:

- (i) to expand communications within the bioscience community;
- (ii) to increase the public understanding and appreciation of the biosciences;
- (iii) to identify, analyse and solve problems relating to research, development and commercialisation of biosciences;

(iv) to provide career support via the BayBio website;
(v) to centralise life sciences information resources.

Built by the founders of biotechnology in the Bay Area, BayBio was founded in 1990 by a consortium of universities, public officials, educators and bioscience executives to foster a regional climate "in which bioscience can continue to flourish". In its early years in Oakland, BayBio was a renowned centre for the bioscience job seeking community; the organization then moved its "information headquarters" to San Francisco in July 1999.

BayBio brings the bioscience community together into one collective group, and provides a forum for its members to convene and interact to exchange information and ideas. BayBio also provides regional visibility for companies through event collaboration and sponsorship, and speaking opportunities. As the only bioscience organisation in Northern California, BayBio serves the entire region's life science companies (nearly 820), a dozen of private research institutions, nine regional universities, and local government agencies.

Entrepreneurial culture

Entrepreneurial culture refers to the scientists' attitude to look not only at the scientific side of researches but also at the commercial exploitation of their results. Even if a significant part of the knowledge produced in universities and research centres can be codified (e.g. patented or published in journals) and broadly diffused and exploited, tacit knowledge, which is produced in research labs, but also "embedded" in learning processes and procedures must be equally considered. Tacit knowledge can be fully exploited only through the creation of a new company. Starting a new venture, however, is a matter of entrepreneurship rather than science or technology, and requires also managerial competences.

In order to help the diffusion of the entrepreneurial culture among researchers, the following practices can be highlighted:

(i) to teach managerial courses to science and technology graduates. Scientists often fail in the right evaluation of the commercial possibilities of the research results. Moreover, they lack the necessary managerial competences in order to define the company strategy in the long term. Teaching managerial courses to life sciences graduates, even if with a low level of specialisation, may make them "familiar" with managerial approach and may help them in understanding and better evaluating the external support by "professional" managers;

(ii) to provide MBA (Master in Business Administration) and related courses dedicated to life sciences. The characteristics of business models of biotech companies are such to require "tailored" management solutions, that may be made available also for scientists;

(iii) to favour entrepreneurial experiences among researchers. For example, this task may be performed with the possibility for academics to try the entrepreneurial adventure with the option, in case of failure, to go back to the previous academic position (the so called "leave of absence");

(iv) a competition for research grants. A competition for research grants may be introduced at national or local level, ranking universities and public research centres by looking at their scientific outputs and also at their attitude to the commercial exploitation of such outputs. This way to evaluate the scientific activity, adding the perspective of commercial exploitation, reaches the twofold objective to reward leading-edge centres and to force them to improve their exploitation mechanisms.

Biocom Courses (San Diego, US)

Biocom currently offers two professional development courses: "Back to Basics" and "From the Laboratory to Leadership: Developing Scientific and Corporate Leaders". "Back to Basics" aims to educate biotech researchers and employees with basic concepts and techniques utilised in the field of biotechnology, combining lectures with hands-on lab activities.

"From the Laboratory to Leadership: Developing Scientific and Corporate Leaders" is designed to train managers in scientific fields with strong business leadership skills. Both lectures and team projects are used to teach communication skills, goal setting, time management and conflict management.

Exist programme (Germany)

The Exist programme is a comprehensive initiative promoted by the German federal government in 1997 and still ongoing. The initiative is directed to high-tech start-ups with four main objectives:

(i) the permanent establishment of a "culture of entrepreneurship" at universities;
(ii) the consistent translation of academic research findings into economic wealth creation;
(iii) the targeted encouragement of the great potential for business ideas and start-up personalities at universities;
(iv) a marked rise in the number of innovative start-ups.

Among the actions carried out in the programme, particular relevance has been put on "entrepreneurial courses". These courses, with class sizes of nearly 20 post-graduate students with both economic and scientific backgrounds, concern the development of real start-up projects. The initiative, which focuses not only on biotechnology but also on IT and engineering sciences, has been scientifically supported by the Fraunhofer Institute. Up till 2002, over 100 applications have been submitted and almost two-thirds of them have been approved for the actual establishment of new companies.

Mechanisms to attract key scientific people

The broadening of the knowledge market and its growing complexity makes the issue of the recruitment of the best researchers more important and highlights the question of the incentive structure for their activity. Whereas in the past much was said about working conditions for scientific researchers (as, for example, the quality of life in cities and suburban areas, the flexible work schedules and the professional status), nowadays more and more often the discussion regards not only the career paths but also the economic incentives offered to researchers.

In order to develop efficient and effective mechanisms to attract key scientific people at universities and research centres located in the cluster, the following practices can be highlighted:

- to support research networks, research organisations and enterprises (including in particular SMEs), in the provision of structured global schemes for the transnational training and mobility of researchers, and the development and transfer of competencies in research including those relating to research management and research ethics;

- to provide the means for research teams of recognised international status to link up, in the context of a well-defined collaborative research project, in order to formulate and implement a structured training programme for researchers in biotechnology. Networks can provide a cohesive but flexible framework for the training and professional development of researchers, especially in the early stages of their research career. Networks also aim to achieve a critical mass of qualified researchers and to contribute to overcoming institutional and disciplinary boundaries, notably through the promotion of multidisciplinary research;
- to support cooperative research projects among actors at international level. A major outcome of such kind of projects (particularly if successful) is the enhancement of the competitive position of the actors involved. In particular, foreign scientists can know more about the faculty and the research activities (and related grant policies) carried out at universities and research centres of the cluster, perhaps eventually considering moving to them;
- to support the "creation" of leading-edge scientific base. The term creation here refers to the leading-edge attribute. Enhancing the specialisation of each university or research centre in a particular scientific or technological field, in which being able to reach the excellence, may, even if indirectly, be a strong mechanism to attract key scientific people worldwide. Indeed, in this case, scientists are attracted by the scientific "curiosity" and by the possibility to carry out the research activity in a state-of-the-art environment.

As in the case of the diffusion of entrepreneurial culture, also concerning this issue, actions can be undertaken by the central government (or by supranational goverment, as in the case of the EU commission). For example, measures to stop brain drain at national or supranational level may be introduced.

Measures of the European Commission to stop brain drain

Based on a deep analysis of career prospects in the EU, the Communication "Researchers in the European Research Area: one profession, multiple careers" identifies factors that impact on the development of careers in R&D, namely training, recruitment methods, employment conditions, evaluation mechanisms, and career advancement. The Communication proposes concrete steps to encourage and structure improved dialogue and information exchange with researchers and to establish a genuinely competitive research labour market at European level. Indeed, in relative terms the EU produces more science graduates (PhDs) than the United States but has fewer researchers (5.36 per thousand of the working population in the EU compared with 8.66 in the USA and 9.72 in Japan). In order to achieve the objective of raising Europe's investment in research to 3% of gross domestic product (GDP), as decided at the Barcelona European Council meeting in March 2002, the EU will need 700,000 extra researchers. There is therefore an urgent need to improve the image of researchers within society, attract more young people to scientific careers and foster researchers' mobility across Europe and back from other regions in the world. There are still some major obstacles to overcome, including in particular difficulties in cross-sector mobility such as moving from university to private business careers, and the additional problems encountered by researchers attempting to embark on careers in universities outside their own countries.

The initiatives set out in the Communication include:

(i) the launch of a "European Researcher's Charter", for the career management of human resources in R&D;
(ii) a "Code of conduct for the recruitment of researchers" at European level;
(iii) the development of a framework for recording and recognising the professional achievements of researchers throughout their careers, including the identification of tools aimed at increasing the transparency of qualifications and competencies acquired in different settings;
(iv) the development of a platform for the social dialogue of researchers;
(v) the designing of appropriate instruments in order to take into account the necessary evolution of the content of research training;
(vi) the development of mechanisms to ensure that doctoral candidates have access to adequate funding and minimum social security benefits.

9.2.3 Industrial Driving Forces

Industrial driving forces can be identified as in Fig. 9.5:

Other than the scientific base, the industrial base is also an absolute prerequisite to build a geographical concentration of new companies. Six factors can be identified as influencing the birth and development of a cluster from the industrial side:

- presence of industrial base;
- existence of success stories in biotech;
- attraction of new sites of other companies;
- integration among industrial actors;
- support to R&D outsourcing processes and industrial spin-offs;
- mechanisms to attract key managerial and commercial people.

The creation of an industrial base (given some exceptions) and the existence of success stories in biotech can be identified as context factors, representing a heritage of the past of the area.

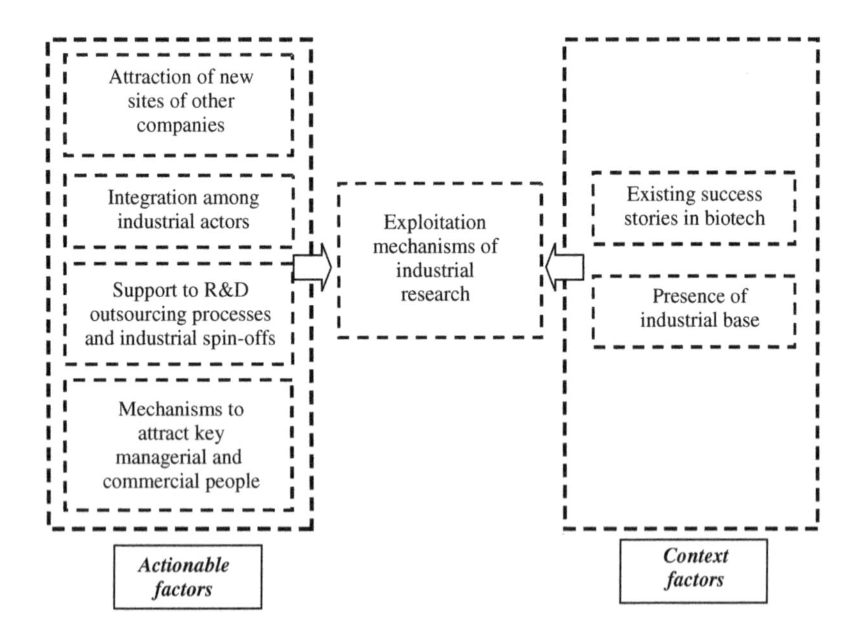

Fig. 9.5: Industrial driving forces.

The other factors, instead, are real actionable driving forces, and also in this case the objective of the related actions should be the creation of a "virtuous" circle.

Attracting new sites and/or key managerial and commercial people allows the increase respectively of the industrial "mass" and of its quality in the area. The same results can be achieved by supporting the R&D outsourcing processes of large companies as well as the spin-off mechanisms. This stimulates the direct creation of new research-based companies as well as the one of service and supplier companies. Finally, benefits of co-localisation can be actually exploited favouring the integration among industrial actors.

Presence of industrial base

The presence of a strong industrial base represents a prerequisite for the growth of a cluster. Particularly, it is true for two reasons: (i) a strong industrial base in the biotech sector represents a "dedicated" market for the research results of the universities and research centres as well as of small DBFs; and (ii) a strong industrial base represents a trigger for the creation of new companies both directly, through the mechanism of industrial spin-offs, and indirectly, favouring the establishment of suppliers and service companies as well as new core biotech companies. For example, in the case of San Diego, the acquisition of the DBF Hybritech by the Eli Lilly forced the scientists of the biotech company, willing to maintain their independence, to create a huge number of new start-ups (more than 20 in the first years after the acquisition). The major outcome of the acquisition was, therefore, the establishment of a biotech-related industrial base, characterised by a strong innovative environment. The restructuring of the operations at Pharmacia in Uppsala that occurred after the merger with Upjohn triggered the creation of new companies, leveraging existing industrial structures and competences. In the case of Biovalley it was the merger in 1996 of Ciba Sandoz into Novartis that triggered the cluster creation. More than 3,000 highly qualified unemployed people forced local public actors to implement the concept of a Biovalley (i.e. a "Silicon Valley" dedicated to biotechnology in Rhine Region) into a concrete initiative.

As for the scientific base, also in this case it is a matter of evidence that the creation of an industrial base is the result of a long process that concern the history of the geographical area in which clusters are growing. Industrial policies at governmental level may help reaching the excellence in a given sector, with focused investment strategies, or help "revitalising" an area after its crisis facilitating the "conversion" of existing industrial infrastructures. However these processes require a long-term time horizon and are assumed as context factors in our analysis.

Some scholars state that it is possible to establish *ex abrupto* an industrial base through the direct "entrepreneurial" intervention of the central government. An example of such kind of interventions can be found in the rapid development of Japan, in the past, and nowadays of China and India. In research driven sectors (not in manufacturing or in the other labour intensive businesses), however, this still appears quite difficult to create. In the biotech sector, two particularly interesting cases of rapid development of the industrial base are Taiwan and Singapore. In both cases, however, governments decided to support initially the establishment of manufacturing centres, which mostly require less scientific competences and can be easily separated by the R&D centres, aiming at creating a favourable industrial background for the further development of local research activities. In Taiwan the support by the government consists in a huge funding activity, which accounted globally for US$5 billion during the period 2000–2005, and in dedicated industrial polices (i.e. tax incentives for new sites). One of the main objectives in the development of Taiwan's strategy is to establish the area as a link in the international community between R&D and the commercialisation of products, acting as a global service and manufacturing centre for biotech. In Singapore, similarly, the evolution of the biotech context is strongly embedded with the activity of local government that has structured a dedicated central agency (EDB, Economic Development Board).

Existence of success stories in biotech

The existence of successful industrial examples represents a strong incentive to found new biotech companies. The presence of such companies, indeed, becomes an effective way to widespread the entrepreneurial culture among scientists, showing them how to create and run a company in the sector. Moreover, it may represent a key driver in localisation choices of large companies.

Genentech, founded in 1976 by the venture capitalist Robert A. Swanson and the biochemist Herbert W. Boyer, resulted the first established firm of modern biotechnology. Genentech started its activities near San Fracisco, and represented a milestone in creating the today's U.S. largest biotech cluster (the Bay Area Cluster) and in inventing the whole new biotech industry, transferring a brilliant scientific discovery in an effective business model.

In the creation of the cluster of Marseilles, Immunotech played a key role. Immunotech, created in 1982 by some researchers of the Marseilles–Luminy Immunology Centre, represented a major event in the French academic and research environment because this type of spin-offs was (at that time) totally uncommon. The venture turned out to be a success and generated a lot of interest and attention both from the business and the academic side.

In the cluster of Heidelberg, and more specifically within the Heidelberg Technology Park, the successful example of Lion Bioscience (currently the world leader in the bioinformatics tools) strongly contributed to the development of the area.

In some cases, even the public support to high potential DBFs in their initial stages has the deliberate objective to create a successful example, thus speeding up the development of the whole sector in the area.

Celltech, for example, based in Slough (UK), was founded in 1980 with the National Enterprise Board (a governmental institution for the industrial policy) as its major shareholder, in order to create the first biotechnology company in the UK capable of exploiting discoveries in the biological sciences at British universities. Celltech had exclusive rights to inventions emerging from MRC's (Medical Research Council) Laboratory of Molecular Biology in Cambridge.

Attraction of new sites of other companies

The establishment of new sites by foreign companies in the geographical area of the cluster enlarges the industrial base. Some remarks, however, concern the typology of the sites. Indeed, only those sites in which research activities are carried out may provide a strong pull effect to the evolution of the cluster, exploiting the scientific base and providing supply-side advantages within the cluster. Manufacturing plants, instead, generate less value-added activities and give a little (or no) contribution to the enhancement of the competitive position of the cluster.

In order to attract quality new sites from other companies, the following practices can be highlighted:

- the provision of economic incentives for the establishment of new sites, creating an annual budget managed by the local government and assigned with regard to the "quality" and the number of new sites;
- the provision of tax credits for research activities in the first years of establishment. Different from direct incentives, tax credits represent a cost reduction for new sites and are available only for research centres. In some cases, tax incentives may be extended for example to venture capitalists, stimulating the investments in biotech within the cluster.

Science and Technology Plan (Taiwan)

Since 1984 Taiwan is developing a Science and Technology Plan focused on biotechnology. One of the main objective of the Plan is to establish Taiwan as a link in the international community of R&D and the commercialisation of products, presenting Taiwan as a service and manufacturing centre for biotech. Five new science-based industrial parks were created, offering favourable rents and shared machineries (conditions extended also to foreign companies). To this aim the government created a fund of €3.1 billion. Another fund (Development Fund) of €57.5 million supports large scale plant development. Up to 2002, 19% of the Development Fund has been used to support large projects of foreign companies and the creation of around 14 new biotech companies.

Moreover, biotech investments have been stimulated by tax incentives: venture capitalists investing in biotech businesses are allowed a 25% tax rebate on capital gain.

Biotech Facilities Tax Credit (Arkansas, US)

Arkansas' General Assembly passed incentives during the 1997 legislative session that are tailored specifically to biotechnology firms. A summary of the incentives includes:

(i) a 5% tax credit on the cost of biotechnology facilities;
(ii) a 30% tax credit on the cost of employee training when conducted through a higher education partnership;
(iii) a 20% credit for the cost of qualified biotech research that exceeds the cost in base year 1996 (i.e. the 1996 tax year). Any unused credit may be carried forward for 14 years after the tax year in which the credit originated. Cumulatively, these credits can create a significant tax savings for companies that invest in the state and reduce the cost of doing business.

Integration among industrial actors

The closer the relationships between the industrial actors, the stronger the impact of the industrial base on the cluster development. Relations are favoured by localisation, but particularly for industrial actors, the process of formal and informal networking should be strengthened with direct actions. The practices analysed regarding the diffusion of the networking culture in the scientific base can be easily adapted to the case of the industrial base. Moreover, the following practices can be highlighted:

- facilitation of networking among industry participants through the establishment of a forum for the dialogue between the industry and local economic development officials and institutions. Such a forum, focused primarily on the problems in linking the industrial actors within the cluster with the local community, may help the cohesion among different actors;
- encouragement of mature companies to partner with and provide space for start-ups. This solution improves the commercial awareness of start-up companies and provide them additional "incubator" style space and business mentoring, allowing large companies with a close relation with the most innovative products and processes.

Biocom (San Diego, US)
In the mid '90s industry leaders in the cluster of San Diego, the third largest cluster in the world, gathered together with a strong commitment to create

an association that would ensure growth and expansion opportunities and represent the industry's interests on a local, state and national level. Biocom was founded in 1995 by the merger of the Biomedical Industry Council (BIC) and the San Diego Biocommerce Association. The organisation was initially created to provide advocacy for industry on local infrastructure issues having an impact on future industry growth. Over the last six years, the Biocom has grown into one of the largest and most acknowledged life science regional trade associations in the nation. Biocom currently operates for members in the areas of public policy advocacy, industry events and conferences, promotion of the industry, professional development programs, industry news and information, and, most importantly, purchasing group and member discounts that substantially affect the bottom line of the companies' value chain.

Support to R&D outsourcing processes and industrial spin-offs

As a consequence of the biotech revolution that changed the way the research activities are carried out in the pharmaceutical sector, most Big Pharmas tend to specialise in development and marketing (where the basic capabilities are substantially unchanged), outsourcing (whole or part of) the research phase. In the latter, indeed, a broader and more complex set of biotech-based technologies and disciplines emerged, thus "eroding" the leadership in innovation of traditional large organisations in favour of the smaller DBFs. Moreover, innovations in the sector are significantly more rapid than in the past and it is much more difficult to keep in-house activities up to speed in all fronts. Restructuring processes of large companies in recent years allowed the birth of many innovative biotech companies through spin-off (or similar) mechanisms. Direct intervention in supporting and favouring this dynamic, which concerns high value-added activities, should be implemented to enhance the generation of new companies within a cluster.

In order to support the creation of new companies through the R&D outsourcing processes, the following practice can be highlighted: favour of MBOs and industrial spin-offs from large corporations. Such mechanisms allow to exploit the industrial research, "re-vitalising" the existing labs and infrastructures (otherwise dismissed) and maintaining a high level of employment.

Biovalley (Germany–France–Switzerland)
The cluster of the Biovalley, centred in the triangle Friburg, Strasbourg and Basel, developed recently under the joined effort of the governments of Germany, France and Switzerland. The idea to support the biotechnological sector was initially conceived by some private German actors at the end of the '80s. Nevertheless, up to 1996, the related initiatives were rather rare. In 1996 the cluster effectively started when the process of M&A leading to the creation of Novartis created nearly 3,000 qualified people unemployed in the life science sector. Based on a joint initiative of the regional and local governments, development agencies, universities and private companies the Biovalley Promotion Team was founded. The Biovalley Promotion Team had the main objective of supporting the creation of a biotech cluster, and primarily the spin-off process and the R&D outsourcing from Novartis. The annual budget of the Biovalley Promotion Team was around €1 million, initially from public funds and subsequently mixed (private and public). The project aimed at creating 400 new DBFs (mainly industrial and academic spin-off) and of 3,000 new jobs on a 10-year horizon. Till now, the initiative gained good results, facilitating the creation of 121 new biotech companies (from 1997 to 2000).

Vicuron Pharmaceuticals (formerly Biosearch Italia)
The birth of Biosearch Italia is heavy linked to the mentioned restructuring processes of large pharmaceutical companies. Up to 1995, it was a leading-edge research centre of the Lepetit Group in Italy, focused on the discovery and production of antibiotics. In those years, the Lepetit Group was acquired by the Marion Merrell Dow, which soon afterwards merged with Hoechst and created the Hoechst Marion Roussel, HMR (now Aventis). As a result of these M&A processes, the Italian research centre became "non-core" and was doomed to dismission. The local management, however, decided in 1996 to start an independent company (Biosearch Italia) and, through a management-buy-out, "acquired" the centre (infrastructures, human resources, patents, ...). HMR favoured the MBO with a two years

research contract of €14 million worth. Biosearch Italia also gained the access to public funds by the MIUR (Italian Ministry for University and Research) for nearly €15 million.

In March 2003, Versicor and Biosearch Italia completed a merger to create an international company (Vicuron Pharmaceuticals) focused on anti-infectives.

Axxam (Italy)

In 1994 Bayer established a research centre in Milano (Italy) for the development of biochemical and biomolecular assays. At the end of nineties, Bayer changed its strategy and carried out a strong focus on pharmaceutical units, planning the closure of the other technology and service development units and the recourse to the outsourcing. In Italy, Bayer favoured the spin-off of Axxam from its centre in Milan, granting the new company a five-year research contract of €29 million.

Mechanisms to attract key managerial and commercial people

Even if R&D actually represents the core activity of DBFs, running a successful company requires broader competences, particularly regarding management and marketing. As regarding the scientific base, indeed, the ability of a cluster to attract key managerial and commercial people strengthens its possibilities of success, making available *in loci* an experienced management team for new and existing companies.

In order to better attract key managerial and commercial people, the following practices can be highlighted:

- promotion of new business ideas to the business community. A rapid and broad delivery of the relevant information concerning the main areas of activities of the cluster and of the most promising business ideas (for example spreading the results of business plan competitions) may convince top-managers to move to the cluster;
- support of the creation of specialised consultancy and service companies. Lawyers, accountants, patent agents, recruitment agents and IT support, dedicated to the biotech, are needed in order to create a "favourable

environment" for the growth of the industrial base and to make the geographic area "suitable" for top-level managers.

9.2.4 Supporting Driving Forces

Supporting driving forces can be identified as in Fig. 9.6:

Supporting driving forces represent the factors who influence the general context of the geographical area of the cluster, creating a favourable "environment" for the growth of the biotech sector. The aspects analysed here concern:

- the legal framework;
- the attractiveness of the area;
- the presence of dedicated support infrastructures;
- the public acceptance of biotech activities;
- the international promotion of the cluster.

The positive presence of such kind of factors in the geographical area of the cluster strongly contributes to trigger and to "spin" the "virtuous" circle analysed with regard to the scientific and industrial base and also facilitates and attracts investments, thus strengthen the whole scheme of driving forces.

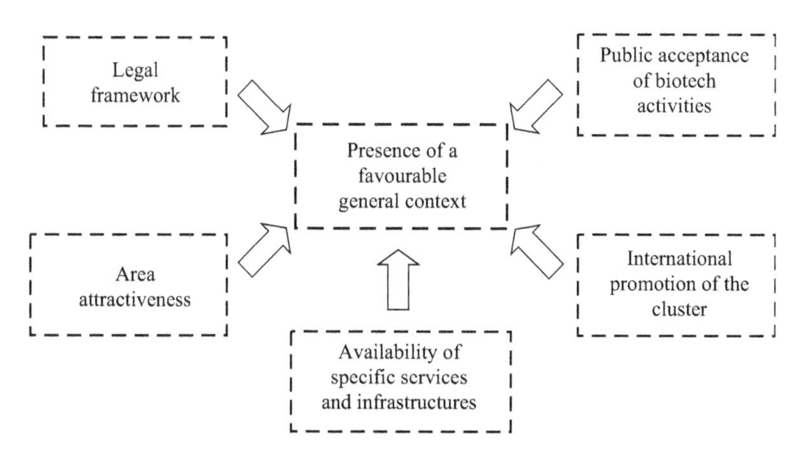

Fig. 9.6: Supporting driving forces.

Legal framework

A well-defined and appropriate legal context is undoubtedly a pre-condition to the development of a science-based, highly innovative sector.

Major issues regarding particularly the biotech sector are: (i) IP rights (i.e. the set of rules which regulates the rights of inventors in exploiting research results); and (ii) bio-security (i.e. the set of rules which regulates the research and production activities, primarily aiming at avoiding risks for workers) and bio-labelling (i.e. the set of rules which regulates the labelling procedures, particularly for food products).

(i) *IP rights*

In the case of academic researchers, particularly the rules regarding the IP rights may represent a major problem in the exploitation of the research results. Concerning this issue, the definition of the legal context is primarily due to the central government; moreover, each university may define internal rules, aiming at supporting (or eventually contrasting) the exploitation of research results. Therefore, the analysis of the legal framework should be conducted at two levels: (i) the national (and/or regional) legal context; and (ii) the local set of rules (i.e. the set of rules within universities and research centres. It is a matter of evidence that the first level of analysis differentiates only clusters in different countries, whereas local rules may explain differences in the development paths of specific areas within the same region.

Concerning the first level, major differences can be found in the US-EU comparison. In US, up to 1980 (Bayh–Dole Act), hundreds of valuable patents were still unused because the government, which sponsored the research that led to the discovery, lacked the resources and links with industry needed for development and marketing of the inventions. Yet the government was unwilling to grant licenses to the private sector. The response to this problem was the reorganisation at central level of the IP rights, enabling small businesses and non-profit organisations, including universities, to retain title materials and products they invent under federal funding. In Europe, instead, patent rules differ in each country (also within the European Union), and the costs and the complexity of the bureaucratic process to extend a patent at the EU level are consistently higher than in US.

(ii) *Bio-security and Bio-labelling*

Bio-security and bio-labelling rules currently have an impact bounded to agbio and nutraceutical applications and do not "interfere" with the development of pharmaceutical applications (that play the primary role in the biotech sector). However, the lack of a clear legal definition, for example, of the use of GMOs, may have a depressive effect on the full exploitation of the scientific innovations related to biotechnology.

Possible support actions are the following:

- to define IP rights of researchers and universities, enabling inventors to retain and exploit commercially, through close relations with the industry, the outcomes (products and technologies) of their researches;
- to speed up the related procedures especially for patent approval;
- to define, particularly in Europe, a common legal framework in order to exploit the advantages of integration among countries.

Bayh–Dole Act and Diamond vs Chakrabarty decision (US)

The US was the first nation to face the problem of IP rights, and already in 1980 precise indications came out both from the Congress and the Supreme Court. Particularly two milestones put the bases for the future growth of the biotech sector: (i) the Bayh–Dole Act, formerly the "Patent and Trademark Act Amendments of 1980"; and (ii) the Diamond vs Chakrabarty decision.

The Bayh–Dole Act created a uniform patent policy among the many federal agencies that fund research. It enabled small businesses and non-profit organisations, including universities, to retain title materials and products they invent under federal funding.

Congress perceived the need for reliable technology transfer mechanisms and for a uniform set of federal rules to make the process work. One major impetus for the bill was the lack of capability of the federal government to transfer technologies for which it had assumed ownership. The few federal agencies that could grant patent title to universities were overregulated, with conflicting licensing and patenting policies. Technology transfer under those conditions was operationally prohibitive for universities and made them reluctant to enter the technology arena. The stability provided by the Act has spurred universities to become involved in transfer of technology from their laboratories to the marketplace. The ability to retain title and to license their inventions have been a healthy incentive for universities. Such incentive is needed, since participation in patent and licensing activities is time-consuming for faculty and must be done in addition to research and teaching priorities. A 1997 survey conducted in US by AUTM (Association of University Technology Management) reports that 70% of the active licenses of responding institutions are in the life sciences, particularly in biotechnology. Most of the inventions involved were the result of federal funding (for example: the Cisplatin and Carboplatin cancer therapeutics, Michigan State University; the Haemophilus B conjugate vaccine, University of Rochester; the Recombinant DNA technology, central to the biotechnology industry, Stanford University and University of California).

In 1980 there were approximately 25–30 universities actively engaged in the patenting and licensing of inventions. It is estimated that there has been close to a ten-fold increase in institutional involvement since then (approximately US$30 billion of economic activity each year, supporting 250,000, jobs can be attributed to the commercialisation of new technologies from academic institutions).

The reorganisation of Intellectual Property rights provided by the Bayh–Dole Act, however, would not have been able alone to starting the biotech sector growth. Another difficult matter was the one concerned with the possible patent objects. Before 1980 the Patent Office was not allowed to grant patents on living organisms, but it is a matter of fact that biotechnology, especially at its origin (with recombinant DNA as main technique), was based on genetically engineered life-forms. The lack of patents' accordance was an high risk for future development. A legal case under the Supreme Court of the United States settled the question.

In 1980 Anand Chakrabarty filed a patent application related to his invention of a human-made, genetically engineered bacterium capable of breaking down crude oil, a property that is possessed by no naturally occurring bacteria. The Patent Office Board denied the application on the ground that living things are not patentable, but the U.S. Supreme Court ruled that the bacterium was eligible for a patent because it had been genetically altered, and was therefore "new, not obvious, not in its natural state, and useful for research".

Genetic Engineering Act (Germany)
Since 1980, German biotech legal context has passed through many phases of innovation. The Genetic Engineering Act is a set of law, amended during the early '90s and aims to clearly define standard legal authorisation of laboratories and production facilities, as well as field trials with genetically modified organisms. Even if quite restrictive, the set of law clearly defined some characteristics of the manufacture activities for therapeutic products deriving from biotech research and field trials of GMOs, permitting anyway the development of these activities. This set of initiatives solved many local problems related to new manufacture facilities authorisations, which strongly affected the development of German biotech context in '80s. In addition to this, the law has a federal effect, making the legal context more uniform, even if some characteristics remained at local level, such as many elements of the approval processes.

Area attractiveness

General infrastructures (transports, ICT infrastructures, ...) and "quality of life" parameters (housing, schools, entertainment, as well as climate and landscape) are key factors to improve the area attractiveness, particularly with regard to human resources. Most of these issues are not "manageable" but some actions may be undertaken in order to:

- improve "family life" services. To attract technical and scientific staff to move from their original regions and also to make available partners in professions such as nursing or teaching (particularly if the cluster is in a region with high housing costs), provision of affordable housing and even low interest loans may be effective;
- to plan adequate industrial spaces and logistics for the growth of the cluster. Expanding industrial areas should maintain a close proximity to the first biotech industrial sites of the cluster, in order to retain the advantages of co-localisation and interaction. Moreover, the identified development areas have to comprehend proper infrastructure and telecommunication capabilities to meet the needs of biotech companies;

- to review the list of land use regulations (i.e. the set of rules which regulates, creating a zone map, the establishment of industrial and residential sites) identifying specific areas for biotech labs and manufacturing facilities in appropriate zone categories, thus minimising the level of discretionary review by local institutions.

Sophia Antipolis (France)

Sophia Antipolis in the Cote d'Azur (France) is Europe's foremost Science Park, with 2,300 hectares, two-thirds of which are landscaped open spaces (destined to expand to the north and through associated sites over time) and a total of 25,900 jobs and more than 2,100 companies. Sophia Antipolis was made official by the Comité Interministériel d'Aménagement du Territoire (Interministerial Committee for Land Development) in April 1972, led by a joint syndicate developer, in 1974, under the name of Symival, which then became Symisa. Symival delegated in the same year the operational workload of Sophia Antipolis to the French Riviera Chamber of Commerce. The Sophia Antipolis Science Park presents a targeted concentration of skill intensive sectors: particularly, information technology (electronics, computing, as well as telecommunications and networks) and life science and fine chemicals (biotechnology, pharmaceuticals, medical imaging, ...). Four types of players are hosted in Sophia Antipolis: (i) major companies and multinationals, (ii) SMEs and start-ups, (iii) public sector research and higher education, and (iv) networks of professional associations, thus enhancing the integration among different actors and the advantages of co-localisation. As for the information technologies, they represent 26% of the companies, 49% of the jobs and 29% of the park's premises, whereas life sciences only account for 4% of firms, 8% of jobs, and occupy 12% of the premises. However, life science companies present the highest growth rate, accordingly with their long-term perspectives.

Sophia Antipolis has never experienced negative variations, not even during international business crisis. The Science Park continues to attract both French and foreign investors and is appreciated as a label of quality by the companies concerned. According to a recent survey of top managers, "Sophia Antipolis is a place where the grey matter likes to settle down to stay and prosper". The 1,500 hectares of greenery planted with Mediterranean species that surround the park make up the "green belt" of Sophia Antipolis and are largely open to the public. Moreover, 150 hectares dedicated to leisure and to residential areas are home to 3,500 families, 70% of which come from outside of Sophia.

Availability of specific services and infrastructures

In order to facilitate the development of an adequate industrial base within the cluster particularly, dedicated infrastructure is needed. Among the usual services the following can be highlighted: (i) incubators, providing spaces and shared service facilities (secretary as well as wet labs) for early-stage start-ups; (ii) science parks, providing analogous infrastructures and services, house and support biotech companies emerging from incubators as well as later stage companies; and (iii) hospitals and clinics, supporting the clinical development phases for biotech drugs and diagnostic technologies.

In order to enhance the availability of specific services and infrastructures, the following actions can be undertaken:

- creation of public biotech-dedicated infrastructures, through direct investments from the local or central government. Public incubators and science parks, as no-profit organisations, represent an effective way for many small start-ups to overcome the lack of capital, reducing initial investment needs;
- promotion of the development of pilot manufacturing and contract manufacturing facilities (bio-processing, fermentation, and cell culture), even "inside" large existing companies. Indeed, it is very difficult, in capital shortage, for new companies to build their own manufacturing facilities;
- support of the creation of corporate incubators. This practice goes further than the previous one. It represents, for large companies, an effective way to "saturate" their plants and infrastructures, to gain alternative revenues and to engage close relations with innovative start-ups. At the same time, start-ups gain the access to state-of-the-art facilities to develop their scientific and business ideas.

Moreover, the presence of specialised service providers is needed too. Lawyers, patent assistants as well as plant engineers in the geographical area of the cluster may facilitate the establishment of new industrial sites, reducing searching costs for new companies.

IZB (Munich, Germany)
IZB is the largest incubator centre in the biotech cluster of Munich. Launched in 1995, the initiative was supported by public actor at local and regional level. The total financing by the Bavarian government was about €45 million. Due also to the winning of the BioRegio initiative the centre became one of the most important actor for the support of first phases of start-up companies in the Munich area. The incubator offers logistic services and some shared machineries. Furthermore the structure hosts Bio$^{\underline{m}}$AG which offers many other services (consultants, financing,...). Because of the great success of the initiative, and with the aim to differentiate the focus of the biotech, a second incubator focusing on green biotech was created. The first site currently hosts 23 enterprises, some of which passed the start-up phase with 2 companies going public in 2002. Seven companies are currently hosted in the new site.

Public acceptance of biotech activities

Public acceptance of biotech activities refers to the positive "feeling" of the social community towards the sector. A favourable social environment may have a strong effect on clusters growth, "spurring on" the workforce and encouraging entrepreneurship in biotech activities. In order to enhance the public acceptance of biotechnology within the cluster, the following actions can be undertaken:

- open debates about benefit and risks of biotechnology promoting mutual understanding between the general public and the scientific community. Social as well as ethical issues should be discussed, avoiding common "prejudices" and adopting a scientific perspective, thus disseminating a correct view of biotech possibilities and risks;
- development of a program for the local biotechnology industry to interact with local high school and community college teachers and students regarding the job potential and opportunities within the cluster. The programme should also allow the development of curricula to meet the basic educational needs of the biotech industry;

- development of seminar programmes and hands-on workshops at school and business sites, eventually encouraging younger employees to go out into the community "to spread the word" on biotechnology activities and career opportunities;
- public promotion of companies and university lab activities, for example increasing the number of business Open Day, thus allowing the public to know what goes on in biotech labs.

Biotechnology in Switzerland

First of all, in Switzerland the use of biotechnology for health-related research and product development enjoys popular backing. In June 1998, the Swiss voted in favour of encouraging research in genetic engineering, demonstrating a diffused acceptance and acknowledgment of biotechnology. Moreover, a strong action to enhance the public acceptance of biotech activities is engaged by the Swiss government. As a service to the Swiss society, two agencies have been created to assess the impact of biotechnology applications, and to promote informed public debates and decision making. Both agencies participate in national and international networks. The Biosafety Research and Assessment of Technology Impacts (BATS) focuses in the scientific safety assessment of transgenic plants and foods derived from genetically modified organisms (GMOs). It also participates in national and international projects on the assessment of social and ethical aspects of technology impacts. The Swiss Agency for Biotechnology Information and Communication (BICS) was instead created to fulfil the public's need for information about this rapidly changing sector, with a specialised library and an information service on biotechnology available online. A website (www.bioweb.org) created in 2000, provides a huge variety of biotech related contents (currently 684 articles, 917 news items, 1,312 links with descriptions, library catalogue with 3,981 entries and an integrated glossary with 425 terms, ...) in English, French and German.

Another important program in supporting biotech acceptance, an generally technological and scientific culture's improvement among people, is "Science et Citè". It is a festival sponsored by the Swiss Government in university cities to promote mutual understanding between the general public and the scientific community. The festival's program comprehends different activities (monthly discussion forums in different cities' Café, a permanent roundtable currently on environmental problems, a newsletter and forum service regarding the theme of stem cell research, ...) and is supported also by big national biotech firms, for example Serono.

Gene Cafes (Evry, France)
The Gene Cafes are quarterly meetings based on the model of "Philosophy Cafes" and take place in different cafes, restaurants near Evry. They offer the public an opportunity to meet and ask questions to the scientists and industrialists working on campus. At each meeting, Pierre Tambourin, Chief Executive of Genopole, hosts one or more speakers. Among the recently debated themes: the correct use of genetics, the benefits and risks of therapeutic clones, the biotechnology impact on the environment.

International promotion of the cluster

To make the cluster known worldwide as a centre of industrial and scientific excellence enhances its attractiveness, especially for qualified human resources. Possible support actions are the following:

- development of a region-wide marketing programme and materials to provide information about the geographic area of the cluster and its resources and pivotal activities to biotech companies interested in relocating to the area;
- global promotion the cluster, through a centralised promotion strategy. It does mean, for example, to attend (as "cluster representatives") dedicated trade fairs and conferences, to publish articles on newspapers and other scientific publications, to set up a website focused on the biotechnology industry within the cluster.

The Scotland promotion scheme
Scotland programmes to internationally promote its biotech activities comprehend: (i) the Biotech Scotland website; (ii) the Bioinformatics Forum; and (iii) the Biotech Talent Scotland website.

The Biotech Scotland website provides the latest news and information from the Scottish biotech community, as well as the latest world-wide biotech news. It is specifically devoted to promote Scottish organisations

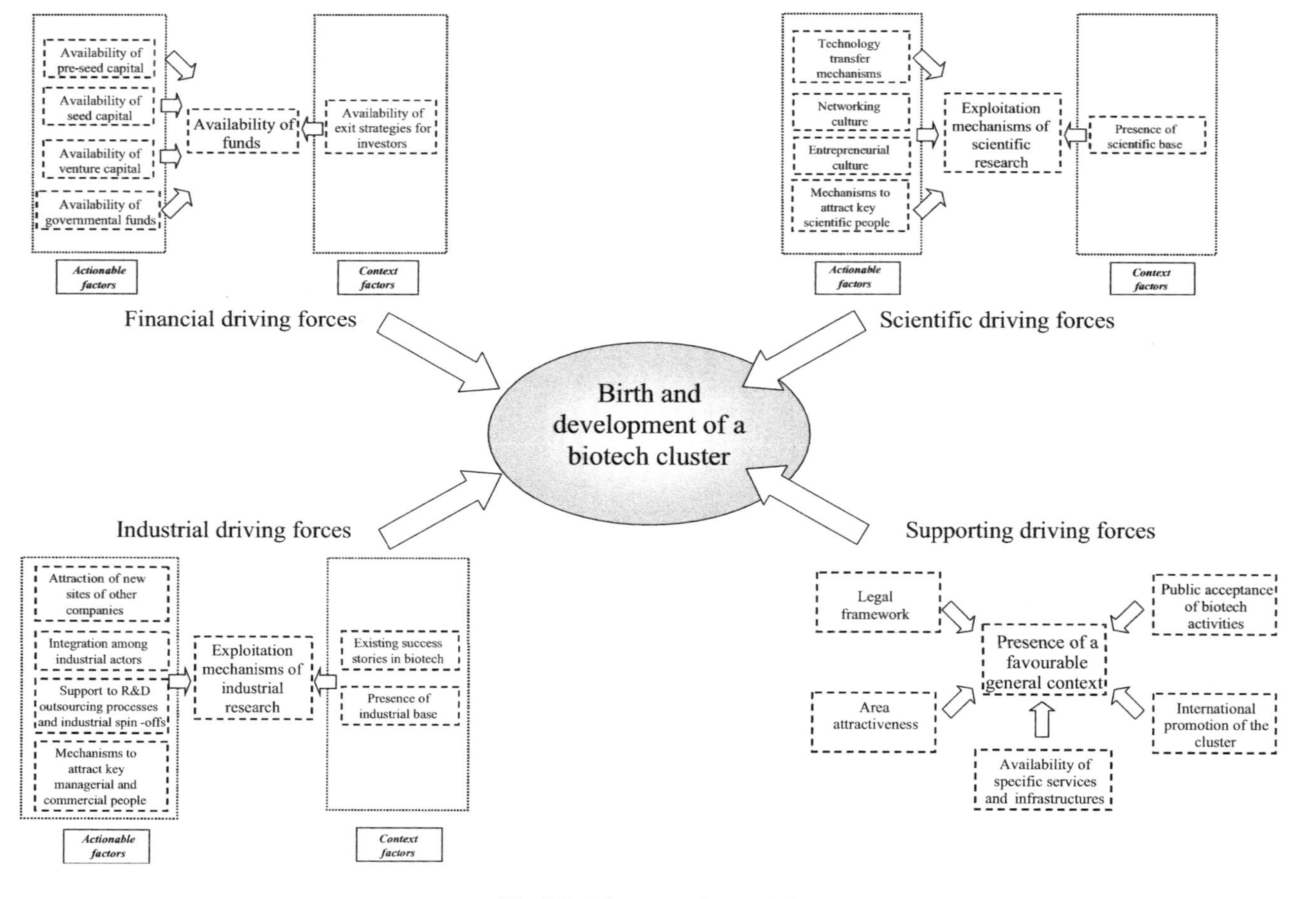

Fig. 9.7: The normative model.

working in the life sciences. Among the services this website offers a keyword searchable databases of all Scottish biotech firms, together with academic and research institute.

As a reflection of both the potential and actual strengths, the bioinformatics community have recently formed the Scottish Bioinformatics Forum. This forum is an informal organisation for researchers primarily in Scotland, whether academic or commercial, and uses as main communication's method the mailing list. It has four key objectives:

(i) to increase the number of effective multi-disciplinary projects being conducted in Scotland;
(ii) to raise both the national and international profile of bioinformatics in Scotland;
(iii) to attract quality people, firms and investments into the Scottish bioinformatics community;
(iv) to provide a stimulating learning environment, leading to improving skills, training and knowledge of bioinformatics amongst science graduates within Scotland.

Finally, the Biotech Talent Scotland website, which was launched in 2003, has the objective of attracting biotech professionals from other countries, by offering a concise and comprehensive view of job opportunities among Scotland's biotech sector.

9.3 The Normative Model

A complete view of the normative model is presented in Fig. 9.7.

10 Conclusions: Forms of Cluster Creation in Biotech

In the previous chapters, we have examined in detail the processes of birth and development of biotech clusters. At the end we present a normative model. The birth and development of a cluster can be seen as a virtuous cycle, where a central role is played by the continuous generation of new science-based companies (Fig. 9.1). A pre-requisite to the birth of a cluster is the presence of a strong scientific and/or industrial biotech base. The process of generation of new companies also requires the availability of funding programmes tailored to the funding of new high-tech ventures. Finally, a fourth factor is given by the characteristics of the general context: the presence of a favourable "environment" (normative, social, historical, and infrastructural context) can actually facilitate the process.

Therefore, four main driving forces can be identified (Fig. 9.2):

- *financial driving forces*, which concern the availability of funds for the biotech companies;
- *scientific driving forces*, which concern the exploitation mechanisms of scientific research;
- *industrial driving forces*, which concern the exploitation mechanisms of industrial research;
- *supporting driving forces*, which concern the presence of a favourable general context.

These elements are relevant in each cluster. However, the process of birth and development of the clusters examined strongly varies from case to case. Two major categories can be identified:

- *spontaneous clusters*, where the cluster has been the result of the spontaneous concentration of the key factors enabling its birth and development;
- *policy driven clusters*, where the trigger was the strong commitment of governmental actors willing to set the conditions for the development of the biotech cluster.

Few other cases cannot be classified among the two categories defined above and are the result of hybrid processes.

10.1 Spontaneous Clusters

Spontaneous clusters are born and grown as the result of the concentration of specific conditions, without the direct commitment of public actors. This model of cluster birth and development took place mostly in US and UK. The usual elements that allowed the cluster to develop and grow are:

- the presence of an excellent scientific base, which is often the result of strong public investments in basic research done in past decades;
- lean exploitation mechanisms of scientific research, especially:
  - technology transfer mechanisms, strongly sustained by initiatives such as industrial liaison offices, technology transfer offices, venture supporting services provided directly by the universities and the research centres;
  - a strong diffusion of the entrepreneurial culture, which also means that among scientists and researchers, there is a strong propensity to commercially exploit the results of their research;
- diffusion of innovative funding mechanisms, which means that there are in place funding schemes (especially related to seed and venture capital) tailored and appropriate for high-tech new ventures;
- the presence of a well defined legal framework (US and UK were the firsts to set up clear laws concerning the scientific research in the biotech sector and to facilitate the industrial exploitation of the research results).

Two main examples of this kind of clusters are the Bay Area and Cambridge.

An interesting aspect of such clusters is that they have not grown around a pivotal organisation: in other words there was not a central organisation that favoured the development of the cluster. These initiatives were undertaken afterwards, such as BayBio in San Francisco and ERBI (East Region Biotechnology Initiative) or EEDA (East Anglia Development Agency) in Cambridge. These organisations have been founded mostly with the aim to favour the connections among the actors in the cluster, conduct lobby actions towards governmental actors, promote the cluster internationally. Another specific factor is that incubators and science parks played a limited role in the development of the cluster. Usually they did not exist at the beginning and their establishment was the result of later initiatives.

10.2 Policy-Driven Clusters

In the case of policy-driven clusters, the actual triggers of the birth of the clusters are the direct actions of policy makers.

Policies can be divided in two categories:

- *industry restructuring policies*, in which the decision of governmental actors to undertake direct actions is the response to an industrial crisis;
- *industry development policies*, in which the direct actions of public actors are the consequence of the decision to foster the biotech sector.

10.2.1 Industry Restructuring Policies

The starting condition is typically the crisis of an industrial sector (or of a single large company) that was providing the strong industrial base to a certain region. In such cases, governmental actors may decide to undertake initiatives to ensure that new jobs are created for redundant people. This is usually done leveraging the existing competencies in the area.

The key driving forces in this case relate to:

- exploitation mechanisms of industrial research, especially favouring: (i) the processes of outsourcing of industrial research to third parties leading to the creation of industrial spin-offs, and (ii) management-buy-outs that allow the local managers to create a new company from the dismission of an existing facility;
- governmental funds dedicated to support the creation of industrial spin-offs.

Cases of this kind are: the cluster of Uppsala, which started as a response to the restructuring of the operations of Pharmacia after the merger with Upjohn; the case of the Biovalley, which was created as a response to the unemployment generated by the merger between Ciba and Sandoz.

Usually these processes are governed by a central actor specifically created to promote and manage the restructuring process: the Stunts Foundation in the Uppsala cluster, the Biovalley AG (formerly Biovalley Promotion Team) in the Biovalley cluster.

10.2.2 Industry Development Policies

Industrial development policies are the result of the deliberate decision of governmental actors to facilitate the development of the biotech sector. Usually the starting condition is the existence of a large and strong scientific base. The intervention of the governmental actor aims to put in place factors enabling the birth and development of an industrial base of biotech firms. The key aspect becomes to improve the entrepreneurial "attitude" and favour the generation of new companies.

The driving forces in such cases are:

- the exploitation mechanisms of scientific research, especially those:
  - favouring the diffusion of entrepreneurial culture, facilitating the creation of new companies;
  - supporting technology transfer mechanisms;

- supporting driving forces, especially those:
  - increasing the availability of infrastructures and services supporting the creation of new companies (especially incubators and science parks);
  - establishing a clear and favourable legal framework, concerning both the legislation about biotech research and the management of IP;
  - favouring the public acceptance of biotech.

The two most important examples are the German and the French cases. In the German case, given that a certain level of activity was already in place, the policy was directly devoted to supporting the foundation of new companies. Infrastructures such as incubators and science parks were already available. Therefore, the choice was to select few areas in the country (through a national competition) and directly fund new companies (only if able to collect the same amount from private investors). In the case of France, the governmental action concentrated on the creation of an infrastructure of technology transfer centres, devoted to promoting entrepreneurship among scientists and researchers, through the provision of funds, space and advice to new companies.

These policies require a central organisation acting as a pivotal actor in the cluster, managing services and funds to new companies. Examples are the Genopoles in the French case, Bio$^{\underline{m}}$AG in the Munich cluster, the Heidelberg Technology Park in Heidelberg.

10.3 Hybrid Clusters

In some cases, the birth of a biotech cluster is the result of hybrid processes. The two major cases are San Diego and Milano. In the case of San Diego there was already a high-tech cluster (focused on ICT) that grew up spontaneously in place. The crisis of the military market brought a strong decline of the cluster, which was converted to biotech through supporting actions of local government. This means that there were in place the factors enabling a high-tech cluster to develop, and the action was directed to the conversion of the industrial base. Several initiatives were created to support the process.

In the case of Milano the governmental actors played a key role supporting the management-buy-outs which were the result of the dismission of facilities by large multinationals. However, the support was not part of a global plan aiming to develop the sector in Italy but simply was given case by case. Therefore the small cluster that is growing up in Milano is the result of the entrepreneurial initiatives of individuals supported by the public actors in the development of their ventures. No central actors play a role in such process.

References and Further Readings

1. Adelberger K.E. (1999) "A Developmental German State? Explaining Growth in German Biotechnology and Venture Capital", BRIE Working Paper, n. 134.
2. Adler A.A., Conklin D. (2001) "Bioinformatics", Encyclopedia of Life Sciences, Macmillan Publishers Ltd., Nature Publishing Group, London, UK.
3. Adler D.A. (2001) "Human genetics: online resources", Encyclopedia of Life Sciences, Macmillan Publishers Ltd., Nature Publishing Group, London, UK.
4. Adomeit A., Baur A., Salfeld R. (2001) "A new model for disease management", The McKinsey Quarterly, n. 4.
5. Alberghina L. (2001) "Genomica, Postgenomica e Biotecnologie di fronte alla complessità biologica", Keiron, n. 7, 64–75.
6. Alberghina L., Cernia E. (1996) "Biotecnologie e Bioindustria", UTET, Torino, Italy.
7. Allansdottir A., Bonaccorsi A., Gambardella A., Mariani M., Orsenigo L., Pammolli F., Riccaboni M. (2002) "Innovation and competitiveness in European biotechnology", Enterprise Papers, n. 7.
8. AltAssets (2002) "Delivering on discovery: Private Equity investing in Biotechnology".
9. Altshuler J., Flanagan A., Steiner M., Tollman P. (2001) "A revolution in R&D — How genomics and genetics are transforming the biopharmaceutical industry", The Boston Consulting Group.
10. Audretsch D.B., Stephan P.E. (1996) "Company-scientist locational links: the case of biotechnology", The American Economic Review, vol. 86, n. 3, 641–652.

11. Audretsch D.B., Stephan P.E. (2001) "Biotechnology in Europe: lessons from the USA", International Journal of Biotechnology, vol. 3, n. 1/2, 168–183.
12. Augen J. (2002) "The evolving role of information technology in the drug discovery process", DDT Drug Discovery today, vol. 7, n. 5, 315–323.
13. Baba Y. (2000) "Development of novel biomedicine based on genome science", European Journal of Pharmaceutical Sciences, vol. 13, n. 1, 3–4.
14. Bastianelli E., Eckhardt J., Teirlynck O. (2001) "Pharma: can the middle hold?", The McKinsey Quarterly, n. 1.
15. Becker M. (2001) "The Biotechnology Revolution", BackOnBiotech.
16. Bergman E., Von Hertog P. (2001) "In pursuit of innovative clusters. Main findings from the OECD cluster focus group", NIS Conference on "Network- and Cluster-oriented policies", Vienna, Austria, 15–16 October.
17. Bergman E., Feser E., Sweeny S. (1996) "Targeting North Carolina Manufacturing: Understanding the State's Economy through Industrial Cluster Analysis", vol. I, Chapel Hill, North Carolina, US.
18. Bhandari M., Garg R., Glassman R. *et al.* (1999) "A genetic revolution in health care", The McKinsey Quarterly, n. 4.
19. Biotechnology Industry Organisation (2001) "Guide to Biotechnology".
20. Biotechnology Industry Organisation (2002) "Guide to Biotechnology".
21. Boschma R. (2001) "Proximity and innovation", Third Congress on "Proximity New Growth and Territories", Paris, France, 13–14 December.
22. Boston Consulting Group (2001) "Assessment of Biotech clusters by the BCG".
23. Breschi S., Lissoni F., Orsenigo L. (2001) "Success and failure in the development of biotechnology clusters: the case of Lombardy", in Fuchs G., "Comparing the Development of Biotechnology Clusters", Harwood Academic Publishers, Chur, Switzerland.
24. Brown S. (2001) "Achieving compliance for biologics", Trends in Biotechnology, vol. 19, n. 8.
25. Bud R. (2001) "History of biotechnology", Encyclopedia of Life Sciences, Macmillan Publisher Ltd., Nature Publishing Group, London, UK.
26. Buffinger N., Mascarenhas D.M. (2001) "Patenting genes and their products", Encyclopedia of Life Sciences, Macmillan Publisher Ltd., Nature Publishing Group, London, UK.
27. Burrill&Company (2000) "Biotech 2000. Life Sciences Changes and Challenges".
28. Burrill&Company (2001) "Biotech 2001. Life sciences: Genomics Proteomics . . . and more".

29. Burrill&Company (2002) "Biotech 2002. Life sciences: Systems Biology".
30. Burrill&Company (2003) "Biotech 2003. Life sciences: Revaulation and Restructuring …".
31. Carroll M.L., Nguyen S.V., Batzer M.A. (2002) "Genome databases", Encyclopedia of Life Sciences, Macmillan Publisher Ltd., Nature Publishing Group, London, UK.
32. Chiaroni D., Chiesa V., Toletti G. (2003) "The biotech revolution in Big Pharma Organization", IEEE-IEMC Conference, Albany, US, 2–4 November.
33. Chiesa V. (2003) "La bioindustria — Strategie Competitive e Organizzazione Industriale nel Settore delle Biotecnologie Farmaceutiche", ETAS, Milano, Italy.
34. Chiesa V. (2001) "R&D Strategy and organisation — Managing technical change in dynamic contexts", Imperial College Press, London, UK.
35. Chiesa V., Toletti G. (2004) "How biotechnology changes pharma R&D: a managerial perspective", International Journal of Biotechnology, vol. 5, n. 2, 125–140.
36. Chiesa V., Toletti G. (1999) "Network of alliances for innovation: the case of biotechnology", Strategic Management Society 19th Annual International Conference in Berlin, Berlin, Germany, 3–6 October.
37. Cook J.D., Miller F.W., Ruiz-Funes J.P.M. (1997) "Food Biotechnology", The McKinsey Quarterly, n. 3.
38. Cooke P. (1994) "The Associational Economy — Firms, Regions, and Innovation", Oxford University Press, Oxford, UK.
39. Cooke P. (1999) "Regional Innovation systems: general findings and some new evidence from biotechnology clusters", Nects-Rictes Conference "Regional Innovation System in Europe", Donostia, Spain, 29 September– 2 October.
40. Cooke P. (2001) " Regional innovation and learning systems, clusters and local and global value chains", Kiel Institute International Workshop on Innovation Clusters and Interregional Competition, Kiel, Germany, 12–13 November.
41. Cooke P. (2001) "Clusters as key determinants of economic growth: the example of biotechnology", in Mariussen A. (editor), "Cluster Policies — Cluster Development? A contribution to the analysis of the new learning economy", Nordregio, Stockholm, Norway.
42. Cookson C. (2002) "Still no end to the slowdown", Financial Times, 29 April.
43. Coriat B., Weinstein O. (2002) "Organizations, firms and institutions in the generation of innovation", Research Policy, vol. 31, 273–290.
44. Dear P.H. (2001) "Genome Mapping", Encyclopedia of Life Sciences, Macmillan Publisher Ltd., Nature Publishing Group, London, UK.

45. DiMasi J.A. (2001) "Tufts Center for the Study of Drug Development Pegs Cost of a New Prescription Medicine at $802 million", Tufts University.
46. Edmunds III R.C., Ma P.C., Tanio C.P. (2001) "Splicing a cost squeeze into the genomics revolution", The McKinsey Quarterly, n. 2.
47. Enright M. (1998) "Regional clusters and firm strategy", in Chandler A.D., Hagstrom P., and Solvell O., "The Dynamic Firm — The Role of Technology, Strategy, Organization, and Regions", Oxford University Press, Oxford, UK.
48. Enright M. (2001) "Regional clusters: what we know and what we should know", Kiel Institute International Workshop on Innovation Clusters and Interregional Competition, Kiel, Germany, 12–13 November.
49. Ernst&Young (1999) "Biotech '99: Bringing the Gap Ernst&Young's 13-th Biotechnology Industry Annual Report".
50. Ernst&Young (2000) "Convergence. The Biotechnology Industry Report".
51. Ernst&Young (2000) "Evolution. Ernst&Young's Seventh Annual European Life Sciences Report".
52. Ernst&Young (2001) "Back to Basics. Ernst&Young's 6-monthly Update of the European Life Sciences Industry H1 2001".
53. Ernst&Young (2001) "Australian Biotechnology Report".
54. Ernst&Young (2001) "Integration. Ernst&Young's Eighth Annual European Life Sciences Report 2001".
55. Ernst&Young (2001) "Millennium in Motion. Global Trends Shaping the Health Sciences Industry".
56. Ernst&Young (2002) "The Year in Review (Europe)".
57. Ernst&Young (2003) "Endurance. The Biotechnology Industry Report".
58. European Commission (1998) "Directive 99/44 EC of the European Parliament and of the Council", Official Journal of the European Communities.
59. European Federation of Pharmaceutical Industries and Associations (EFPIA), (2003) "Efpia Infigures 2002".
60. Flattmann G.J., Kaplan J.M. (2001) "Patenting expressed sequence tags and single nucleotide polymorphisms", Nature Biotechnology, vol. 19, n. 7, 683–684.
61. Garner J.L., Nam J., Ottoo R.E. (2000) "Determinants of corporate growth opportunities of emerging firms", Journal of Economics and Business, vol. 54, 73–93.
62. George G., Zahra S.A., Wheatley K.K., Khan R. (2001) "The effects of alliance portfolio characteristics and absorptive capacity on performance. A study of biotechnology firms", The Journal of High Technology Management Research, vol. 12, 205–226.

63. Giescke S. (1999) "Moving goose ahead: catch up strategies of the German system of innovation. The case of biotechnology", European Meeting on Applied Evolutionary Economics, Grenoble, France, 7–9 June.
64. Gilsing V.A., Roelandt T.J.A., Van Sinderen J. (2000) "New policies for the New Economy cluster-based innovation policy: international experiences", Annual EUNIP Conference, Tilburg, Netherlands, 7–9 December.
65. Goldman Sachs (2001) "Healthcare: Biotechnology. Europe", Goldman Sachs global equity research.
66. Guardia M.J. (2002) "Modelling and control in bioprocesses", Trends in Biotechnology, vol. 20.
67. Hung L.W., Kim S.H. (2001) "Genome, proteome, and the quest for a full structure-function description of an organism", Encyclopedia of Life Sciences, Macmillan Publisher Ltd., Nature Publishing Group, London, UK.
68. IBM Business Consulting Services (2003) "Pharma 2005: Marketing to the Individual".
69. IBM Business Consulting Services (2003) "Pharma 2010: The Threshold of Innovation".
70. Institute of Professional Representatives before the European Patent Office (2001) "An Introduction to Patents in Europe".
71. Jacobs D., DeMan A.P. (1996) "Clusters, Industrial Policy and Firm Strategy: a Menu Approach", Technology Analysis and Strategic Management, vol. 8, n. 4, 425–437.
72. Jain V., Mitchell J.J., Ross P. (2002) "Financing in a down market", Biotechnology Investors' Forum Worldwide, vol. 2.
73. Kluge J., Meffert J., Stein L. (2000) "The German road to innovation", The McKinsey Quarterly, n. 2.
74. La Montagne J.R. (2001) "Biotechnology and research: promise and problems", The Lancet, vol. 358, n. 9294, 1723–1724.
75. Lemariè S., Mangematin V., Torre A. (2000) "Is the creation and development of biotech SMEs localized? Conclusions drawn from the French case", Small Business Economics, vol. 17, n. 1/2, 61–76.
76. Lindgren T. (2000) "Sweden: Strong and Growing", Cap Gemini Ernst & Young.
77. Lord Sainsbury, Minister for Science (1999) "Biotechnology Clusters", UK Counsel.
78. Love J. (2000) "How much does it cost to develop a new drug", Geneva Meeting of the MSE Working Group on R&D, Geneva, Switzerland, 2 April.
79. Mallik A., Pinkus G.S., Sheffer S. (2002) "Biopharma's capacity crunch", The McKinsey Quarterly, Special Edition: Risk and resilience.

80. McCain L. (2002) "Informing technology policy decisions: the US Human Genome Project's ethical, legal and social implications programs as a critical case", Technology in Society, vol. 24, n. 1/2, 111–132.
81. McCutchen Jr W.W., Swamidass P.M. (2001) "Effect of R&D expenditures and funding strategies on the market value of biotech firm", Journal of Engineering and Technology Management, vol. 12, 287–299.
82. Mytelka L., Farinelli F. (2000) "Local Clusters, Innovation Systems and Sustained Competitiveness", United Nations University, Institute for New Technologies, Discussion Paper, vol. 5.
83. Paisner G. (2001) "Europe's biotech VCs are waiting out the downturn by making larger investments", Red Herring, December.
84. Pharmaceutical Research and Manufacturers of America (PhRMA) (2001) "Pharmaceutical Industry Profile 2001".
85. Pharmaceutical Research and Manufacturers of America (PhRMA) (2002) "Pharmaceutical Industry Profile 2002".
86. Porter M. (1990) "The Competitive Advantage of Nations", Mcmillian Publisher Ltd., Nature Publishing Group, London, UK.
87. Porter M. (1998) "The Adam Smith address: location, clusters, and the "new" microeconomics of competition", Business Economics, vol. 23, n. 1.
88. PriceWaterhouseCoopers (1998) "Pharma 2005. An Industrial Revolution in R&D".
89. PriceWaterhouseCoopers (1999) "Pharma 2005. Silicon Rally: The race to e-R&D".
90. PriceWaterhouseCoopers (2000) "Global Pharmaceutical Company Partnering Capabilities Survey".
91. PriceWaterhouseCoopers (2000) "Global Pharmaceuticals Sector Insights. Analysis and Opinions on Merger and Acquisition Activity".
92. Rosenfeld Stuart A. (1996) "Overachievers, Business Clusters that Work: Prospects for Regional Development", Regional Technology Strategies, Chapel Hill, North Carolina, US.
93. Rosenfeld Stuart A. (1997) "Bringing Business Clusters into the Mainstream of Economic Development", European Planning Studies, vol. 5, n.1, 1–23.
94. Sauter T. (2002) "New trends in applied microbiology", Trends in Biotechnology, vol. 20, n. 1, 43–44.
95. Signals Magazine (2003) "The Big Spin (Off)".
96. Swain B.J. (1998) "Biotechnology patents in the pharmaceutical industry", Financial Times, Healthcare Management Reports.

97. Swann G.M.P., Prevezer M., Stout D. (1998) "The Dynamics of Industrial Clustering", Oxford University Press, Oxford, UK.
98. Vizirianakis I.S. (2002) "Pharmaceutical education in the wake of genomic technologies for drug development and personalized medicine", European Journal of Pharmaceutical Sciences, vol. 15, n. 3, 243–250.

Printed in the United States
By Bookmasters